T0076745

GAIA'S WEB

GAIA'S WEB

How Digital Environmentalism Can Combat Climate Change, Restore Biodiversity, Cultivate Empathy, and Regenerate the Earth

KAREN BAKKER

The MIT Press
Cambridge, Massachusetts
London, England

The MIT Press would like to thank the anonymous peer reviewers who provided comments on drafts of this book. The generous work of academic experts is essential for establishing the authority and quality of our publications. We acknowledge with gratitude the contributions of these otherwise uncredited readers.

This book was set in Adobe Garamond Pro by New Best-set Typesetters Ltd. Printed and bound in the United States of America.

Library of Congress Cataloging-in-Publication Data

Names: Bakker, Karen J., author.
Title: Gaia's web : how digital environmentalism can combat climate change, restore biodiversity, cultivate empathy, and regenerate the Earth / Karen Bakker.
Description: Cambridge, Massachusetts : The MIT Press, 2024. | Includes bibliographical references and index.
Identifiers: LCCN 2023029166 (print) | LCCN 2023029167 (ebook) | ISBN 9780262048750 (hardcover) | ISBN 9780262377690 (epub) | ISBN 9780262377683 (pdf)
Subjects: LCSH: Environmentalism—Effect of technological innovations on. | Information society—Environmental aspects. | Information technology—Environmental aspects. | Green technology.
Classification: LCC GE195 .B354 2024 (print) | LCC GE195 (ebook) | DDC 363.7/06—dc23/eng/20231016
LC record available at https://lccn.loc.gov/2023029166
LC ebook record available at https://lccn.loc.gov/2023029167

10 9 8 7 6 5 4 3 2 1

For Philippe, Solenne, and Pauline

Contents

I REGENERATING

All that you touch
You Change.

All that you Change
Changes you.
—Octavia Butler[1]

1 WIRING GAIA

In the summer of 2018, a rare birth caught the world's attention. The baby was born to a young mother named Tahlequah, into a pod of endangered orcas living off the coast of Seattle. Her family had not welcomed a new child for more than three years, and a calf was cause for much celebration. But the newborn stopped breathing less than an hour after her birth. The night the baby died, a volunteer orca watcher spotted members of the pod near the shores of San Juan Island. She later wrote:

> At sunset, a group of five to six females gathered at the mouth of the cove in a close, tight-knit circle, staying at the surface in a harmonious circular motion for nearly two hours. As the light dimmed, I was able to watch them continue what seemed to be a ritual or ceremony. They stayed directly centered in the moonbeam, even as it moved.

When orcas die, the sea usually takes them quickly. But after the ceremony, rather than let her baby's body drift to the ocean floor, Tahlequah cradled her dead child with her head and swam out into ocean, under the shadow of the Coast Mountains, carrying her calf on a funeral procession through the far reaches of the Salish Sea. Orca babies, born after nearly two years of pregnancy, average over eight feet long and can weigh several hundred pounds. When Tahlequah grew tired and her breathing became labored, her relatives took turns carrying the corpse. The procession continued for days, then weeks, and with Tahlequah becoming noticeably thinner, scientists began to worry. As one of the only young healthy females remaining, Tahlequah's survival was key to the pod's future. Yet she refused to let her baby go.[1]

While scientists debated the question of whether orcas can grieve, the public mourned alongside Tahlequah, whose pilgrimage was the most-read story in the *Seattle Times* that year. Washington governor Jay Inslee declared, "I believe we have orcas in our soul in this state." Millions of people around the world followed her journey until seventeen days and over a thousand miles later, Tahlequah released her dead baby to the waters.

Soon after, the Canadian government approved the expansion of a new oil pipeline that would increase tanker traffic and the risk of oil spills in the orcas' primary habitat. Some new safeguards were also introduced, including the restriction of the commercial salmon fishery, and a ban on boats approaching the orcas too closely. But scientists warned this would not be enough in the race against time to save Tahlequah and her last remaining relatives. Her family had been declared one of the seven most "imperiled of the imperiled" species under the Endangered Species Act in 2015. When Tahlequah lost her child, fewer than two dozen of her family members remained.

Tahlequah's baby died before her mother could teach her about her home. The Salish Sea is a place where giants still roam the Earth. North of Seattle, the flanks of the mountains rise to over 10,000 feet, luring rainfall out of dense clouds blowing in from the Pacific. Below the surface lies one of the richest inland seas in the world, where the world's largest octopuses live alongside the world's largest anemones and the world's largest barnacles. The mountains are interrupted only by the delta of the Fraser River, the longest major undammed river south of the Arctic, fed by the largest temperate-latitude ice fields on the planet. Every year, salmon fight their way hundreds of miles upriver to reach their spawning grounds, where they give birth and die. Carried into the forest by bears and eagles, their decaying bodies nourish western red cedars that grow to hundreds of feet and live for thousands of years. The largest trees grow closest to the streams, shading the newborn salmon from the sun. When ready, the hatchlings swim down to the ocean, the sockeye with their gold-colored eyes, the chum with their silver tails, the coho with their white mouths, the pinks with their black-spotted backs, and the mightiest of all: Chinook salmon, the largest in the Pacific, known to the Nuu-chah-nulth as Tyee: chief, champion, king.

Chinook are the preferred, almost sole food consumed by the Salish Sea orcas. A full-grown *Orcinus orca* can weigh over ten tons and grow to the size of a school bus, with a brain size second only to that of sperm whales. To survive, an adult orca needs to consume hundreds of pounds of nutrient-rich Chinook each day. Yet the Chinook are now diminished; although they once ranged from California to Alaska, their territory has shrunk to a few rivers, and they are far less abundant. Their lifespans are significantly shorter than that of their grandparents, and their size has dwindled. The world's largest ever Chinook, caught near Petersburg, Alaska, in 1949, weighed over 120 pounds; these days, its Salish Sea descendants average just over 20 pounds.

The Chinook once allowed Tahlequah's ancestors—who would otherwise have ranged across larger expanses of ocean—to become long-term residents of the sheltered Salish Sea. There, the orcas evolved close-knit, stable, and highly social family groups, which live together for their entire lives, roughly between fifty and eighty years. Led by matriarchs who are often grandmothers or great-grandmothers, the pods evolved complex forms of communication and culture, embodied in unique dialects that young orcas learn from their elders. When the pods fan out for fishing or foraging expeditions, the orcas mark their reunions with greeting ceremonies. Mothers teach their pod songs to their young, songs that can be heard across the Salish Sea.

Like other cetaceans, orcas vocalize in distinct repertoires, unique to their family groups. The Coast Salish tribes know that the songs carry stories, narrated by the orcas as rulers of their domain, carriers of undersea laws within the marine world, nonhuman persons with complex culture and intergenerational relationships. The Lhaq'temish Nation's name for the orcas (*Qwe lhol mech ten*) means "Our relations under the waves." Says Jay Julius, "We've fished alongside them since time immemorial. They live for the same thing we live for: family."

TRAIL OF TEARS

The moment of her baby's death was Tahlequah's second brush with celebrity status. Three years earlier, she had been adopted by Malia Obama in a naming ceremony sponsored by a local museum. Malia and her adopted orca share the same birth year: 1998. For the adoption ceremony, a new name was chosen for the orca: no longer known as J35, she was renamed Tahlequah.

The name was an apt choice. The city of Tahlequah, nestled in the foothills of the Ozarks, was established as the capital of the Cherokee Nation in 1839 after the tribe was forcibly removed from its traditional territory further east. Despite a Supreme Court case that supported their cause, the Cherokee were violently evicted and forced to march over 5,000 miles west along what came to be known as the Trail of Tears, one of a series of genocidal relocations of Indigenous peoples.

Tahlequah's family has followed its own trail of tears. The last whaling station at Coal Harbour, which operated out of an old air force base, closed in 1967. Colonial settler fishermen viewed orcas as a threat. *Orcinus orca* means "belonging to Orcus," god of the underworld. Killer whales, as they were once commonly known, had a fearsome reputation as savage predators of fish, seals, sea lions, and even the calves of larger whale species. Fishermen took pot shots at orcas from their boats. In the early 1960s, Canadian government officials mounted a .50 caliber Browning heavy machine gun on a strategic lookout at Seymour Narrows with the sole aim of slaughtering orcas, which were viewed as pests by local fishermen (fortunately, the officials never fired the gun, for fear of locals being killed by bullets ricocheting off the rock). Even the director of the Vancouver Aquarium had no qualms about asking a local sculptor and part-time fisherman, Sam Burich, to harpoon and kill an orca for use as a model for display.

One fine midsummer morning in 1964, Burich borrowed a harpoon from the whalers at Coal Harbour and set out on an orca hunt. The trip did not go as planned. After he spotted a young bull orca off the waters of Saturna Island, Burich harpooned the animal, but he botched the job. Rather than fleeing, the orca's podmates remained, raising their injured podmate to the surface to breathe. Burich, profoundly affected by the creature's cries and the distress of the pod, could not bring himself to kill the orca. He towed the injured animal back to Vancouver. There, the young orca was confined to a makeshift pen fashioned inside a derelict wharf, and christened Moby Doll. Burich was appointed to be his guardian, spending hours playing music to alleviate the orca's loneliness; at times, Moby Doll seemed to respond, vocalizing in a duet alongside Burich's whistle.

After the aquarium put an ad in the local paper inviting the public to see the savage killer up close, Moby Doll attracted crowds. International attention soon followed. The director of Marineland Studios flew in to offer today's equivalent of $200,000 for the orca, the first ever held in captivity. Marineland had deep pockets: founded by grandsons of Leo Tolstoy and Cornelius Vanderbilt, it attracted celebrity authors and Hollywood starlets filming movies like *Creature from the Black Lagoon*.

It took the orca several months to die. By that time, live orca capture had become big business. Seattle Marine Aquarium owner Ted Griffin borrowed a harpoon gun from the Department of Fisheries and began scouring the waters off the coast, sometimes flying in a police helicopter. Word soon spread of Griffin's desire, and the next time an orca was accidentally netted, the fishermen called before shooting. Griffin had his prize: a promising bull orca, the first to be successfully caught and held in captivity as a tourist attraction. Namu (named after a nearby fishing port) generated a media frenzy, crowds of thousands, and a feature film by the producers of *Flipper*. In the years that followed, orca capture became a multimillion-dollar business, and Griffin a primary supplier. Griffin sold his next orca—the original Shamu—to Sea World. In 1970, he netted eighty killer orcas in one operation in Puget Sound; although not well understood at the time, this was nearly the entire population of the Southern Resident orcas. Releasing the largest adults—who were considered too large and aggressive to safely transport—Griffin's team enclosed the calves in smaller pens. While the babies swam anxiously in the hunters' nets, the adults remained nearby, singing their pod songs. The calves were shipped to amusement parks around the United States. Some remain there today. Lhaq'temish poet and National Geographic explorer Rena Priest recounts how, fifty years later, some of the captive orcas still remember and sing their family songs.[2]

As the number of orcas dwindled, the hunters resorted to more aggressive methods. Fishing vessels, speedboats, and spotter planes were used to track and herd the orcas to designated spots, where the adults were separated from the babies. But when the hunters began to use explosives, dynamiting the fleeing orcas in an attempt to kill the remaining adults and corner the remaining children for capture, a public outcry ensued. In their defense, the hunters argued that the orcas, a mere marine pest, were sufficiently abundant to justify their methods. This claim turned out to be untrue. Through a painstaking manual survey which took months, Michael Bigg, a fisheries scientist, calculated that only seventy-one Southern Resident orcas remained. But scientific proof of the near-extermination of the orcas came nearly too late. By the time a capture ban was fully implemented in the 1980s, only a few dozen orcas—including Tahlequah's grandparents—remained.

Despite the hunting ban, the Salish Sea orcas failed to recover. Fifty years later, Tahlequah's baby was born into a community on the brink of extinction. Her family is one of the most contaminated on Earth: bioaccumulated toxins reduce their fertility, suppress their immune systems, and increase their susceptibility to bacterial infections and parasites. Three-quarters of pregnancies end in miscarriage. Overfishing of the orca's preferred food source, Chinook salmon, has led to undernutrition. Scientists have resorted to buying salmon and dropping them overboard to save the most emaciated pod members from dying of starvation.

Perhaps the most insidious threat to the orcas is marine noise. Toothed cetaceans, like orcas, use sound in complex ways. Their vocalizations may be used to identify individuals, negotiate social relationships, and exchange information. Each pod has its own complex and unique dialect, which orca calves learn from their elders. The pods, led by matriarchs, live together for their entire lives; their dialects form part of their cultural identity, and orcas with different dialects rarely intermingle for long.

Orcas also use sound to sense their surroundings through echolocation, which functions like a personal ultrasound. Echolocation, or biosonar, enables orcas to visualize their surroundings by emitting high-frequency sound waves and listening for echoes, enabling orcas to visualize their environment, find prey, locate kin, and avoid threats. Whereas orcas can see only up to a few meters away, they can hear their surroundings across many miles. Orcas continuously scan their environment with sound, just as humans visually scan their surroundings. Noise pollution thus affects orcas much like smoke from forest fires affects humans; it is profoundly disorienting. Recent scientific studies have suggested that noise may be interfering with the ability of the orcas to hunt Chinook salmon, as they can no longer discern their prey through the onslaught of human-generated noise.

Unfortunately for Tahlequah and her family, the Salish Sea is one of the noisiest marine zones in the world. Shipping noise has doubled during each of the past two decades. Ferries, cruise ships, pleasure boats, and jet skis share the waters with oil tankers, cargo ships, tugboats, dredgers, aircraft destroyers, submarines, and hydroplanes. Because ship noise propagates so well underwater, orcas are bombarded with sound from all directions, creating an acoustic smog; the cumulative noise is both deafening and blinding for the orcas, interfering with their ability to find food, mate, and socialize. Some whales also flee loud sounds by spending more time at the surface and swimming away, making them more vulnerable to ship strikes. Scientists refer to these impacts with terms like "auditory masking," but in layperson's terms the impacts are simpler and starker: orcas are lonelier, more stressed, more disoriented, and more malnourished in a noisy ocean. Oceanographer David Barclay offers a simple analogy: if orcas are trying to look for fish while swimming toward a ship, it would be like looking into the bright sun while trying to spot a bird. Humans no longer kill orcas with bullets and harpoons. Instead, we are serenading them to death.

DIGITAL ORCAS

Amid intense debates about how to save the orcas, a novel solution has emerged that advocates believe could help solve several long-standing challenges facing the orca conservation agenda: digital technology. For example, collaboration between the shipping industry and conservationists has been hampered by the fact that, until recently, relatively little was known about where the orcas spend most of their time. Shipping companies chafe against requirements that their ships slow down or move aside, and protest "just in case" conservation measures that purport to protect whales, even when there may not actually be any whales in the vicinity. To address this issue, scientists have developed a digital app called WhaleReport, which sends real-time notifications of whale sightings to all commercial shipping vessels within ten nautical miles. The whales are detected by an underwater system of digital hydrophones—waterproof microphones that allow scientists to monitor the orcas continuously, even while deep below the surface. The data is then analyzed with the help of artificial intelligence algorithms, cross-checked with sightings from human volunteers, and transmitted to ships' captains. In addition to tracking the location of orcas in real time, the artificial intelligence algorithms can also predict where orcas will be within the next few hours. The hoped-for result: ships will slow down and steer away from orcas, avoiding ship strikes.[3]

Digital technology is also being used to address the issue of noise pollution. On the Canadian side of the border, the Vancouver Port Authority has introduced a scheme to reduce ship noise: quieter ships get a discount on port fees. The easiest way to quiet a ship is to slow it down; currently, the slowdown season currently starts as soon as the whales appear in the inland waters near the port—usually around June—and ends when two weeks have passed without a whale being detected—usually in October. Using digital monitors that function much like speed traps on highways, this program reduces shipping noise when whales are most likely to be in the area, at the height of the salmon migration, enabling the orcas to forage and feed more effectively.[4] A similar program, called Quiet Sound, has been launched on the Seattle side of the border in partnerships with the US Navy, which has launched a challenge calling for a digital ship-mounted sensing device that can autonomously detect the presence of orcas—a dashcam for whales.

These efforts are linked to an ambitious digital ocean monitoring initiative, called Smart Oceans, which is wiring up the oceans along the northwest Pacific coast.[5] Digital tags placed on whales have revealed specific differences in how males and females hunt, and underscored the vulnerability of their echolocation-based hunting system to ship noise. Scientists are using digital sensors called biologgers to track the movements of salmon, the orca's preferred prey. They have also attached biologgers to harbor seals, to find out how many juvenile salmon they are consuming before the orcas get their chance. Below the surface, underwater cables crisscross the ocean floor and digital gliders roam freely, monitoring temperature and currents, acidity and salinity, the movements of plankton and fish, and the subtle seismic tremors that ripple up through the water from deep in the Earth. Microsoft and Google have built artificial intelligence algorithms that automatically analyze noise pollution and orca health, and developed OrcaHello, a system that uses artificial intelligence algorithms to analyze underwater recordings to classify orca sounds and decode their meanings, from pulsed hunting clicks to extended greeting calls.[6] Tahlequah's family is now subject to unprecedented, continuous digital surveillance—every move is tracked, and every sound is recorded. Architects of the Smart Oceans network, which is now the largest of its kind in the world, hope the information gleaned from this monitoring will help scientists reverse the orca population's decline.

GAIA'S WEB

The digitization of the Salish Sea is a microcosm of a global trend. Around the world, technologists are developing ambitious digital environmental monitoring systems. Some of these Digital Earth systems are as small as a pond, while others encompass entire continents. When deployed, they reveal much that was once hidden about the natural world. These digital systems rely an astounding range of hardware, from sensors to drones to satellites.[7] Digital tracking devices are affixed to the tiniest of insects. Honeybee sensors, for example, can measure location, temperature, humidity, or light intensity; some bee trackers function much like the digital chip in your credit card, so that the bees can scan themselves every time they enter or leave the hive.[8] The list of sensors available for whales alone is dizzying in its variety. Some sensors record whale communication, enabling scientists to hear even the tiniest whispers made by mother whales to their babies. Larger sensors record whale movements, sensing pitch, roll, direction, and depth, and revealing previously unsuspected behaviors, like the recently discovered fact that Cuvier's beaked whales can dive to depths of over 6,000 feet, four times the height of the Empire State Building.[9] Satellites and underwater drones monitor the ocean waters in which the whales swim: wind and waves, currents and eddies, temperature and salinity.

Digital Earth devices also monitor from afar. The entirety of the Earth's surface, from the depths of the ocean to the upper atmosphere, is continuously tracked by remote monitoring systems. This creates a "digital twin" of our planet: a virtual model of our world, rendered in digital data.[10] One new Earth imaging system, ICARUS (International Cooperation for Animal Research Using Space) uses satellite monitoring to create a living, real-time map of Earth's animals. By simultaneously monitoring hundreds of thousands of animals to which biochip tracking devices have been attached, scientists can observe their real-time movements from space. Scientists are using ICARUS to document and explain the extraordinary behaviors of animals, as well as to also forecast droughts, floods, volcanic eruptions, and even emerging pandemics.[11] Thousands of systems like ICARUS now exist, monitoring forests and tundra, cities and clouds, and even the evanescent flows of atmospheric gases.

The best way to appreciate the global scale of planetary computation is by way of analogy. Think of the Internet of Things—the sensors embedded in our homes, smartphones, and vehicles. Now think of extending this network to the rest of our planet: an Internet of Earthlings. This network connects digital and ecological systems, enabling profound and powerful new insights about our planet, and also creating novel risks to humans and nonhumans alike.

I refer to these interconnected digital and natural networks as Gaia's Web, and in this book, our primary goal will be to explore how Gaia's Web will impact biodiversity conservation and environmental governance. But before we dive in, a brief intellectual history will be helpful. The concept of Gaia was developed in the late twentieth century through a collaboration between biologist Lynn Margulis and chemist James Lovelock, who used this term to describe the dynamic interactions between the living and the nonliving parts of our planet. Their metaphor, which invokes the Greek goddess of the Earth, crystallizes a deceptively simple insight: biological and nonbiological aspects of the Earth system influence one another. A beautiful example is our atmosphere: in a mutually beneficial exchange, plants breathe in the carbon dioxide that animals breathe out, and animals reciprocate by breathing in the oxygen that plants exhale. Living creatures also influence the weather. In the Amazon, trees release enough moisture through photosynthesis to create low-lying clouds, summoning their own rain.[12]

These examples illustrate the intricate feedback loops between living organisms and their environment through exchanges of energy and elements—carbon, nitrogen, oxygen, phosphorus, and sulfur—in complex biogeochemical cycles. From animals and insects to plants and bacteria, and oceans and atmosphere: the Earth is a system of mutual exchanges between living beings and their environments.[13] The Gaia metaphor captures the fact that these relationships, which emerged over eons, are fundamental to life as we know it, and are—as far as we know—unique in the universe.[14]

When Margulis and Lovelock originally developed their idea, they argued that Gaia was both biological and geological. But Gaia is, as I argue in this book, also increasingly digital. My metaphor of Gaia's Web refers to the interplay of digital and ecological networks, the fusion of the World Wide Web and the Web of Life.[15]

Networks are, of course, widespread in nature. Fungal networks associate with the roots of trees in an intricate "Wood Wide Web" through which different species share food and information, conveyed through biochemical messages.[16] Nutrients cycle from rocks to plants to animals to soil. But digital and ecological networks are not equivalent. Digital devices and networks create waste and pollution rather than ecologically useful nutrients; unlike living systems, they do not reabsorb and repurpose their own biophysical waste. Digital networks do not participate in the cycles of the Web of Life; rather, they are a parasitical machine. The global internet now uses more electricity, and hence generates more climate emissions, than most countries in the world. Scaling up Digital Earth accelerates the tech sector's demand for rare earth minerals, extracted from the earth at great environmental and social cost. Given that the internet is the largest machine humanity has ever built, these digital networks have environmental implications such as increased energy and water use, and growing e-waste. In chapters that follow, I explore the transformative power of digital technology as it is interwoven with the intricate, dynamic ecological relationships that bind us all together on Earth. But this technology is not being deployed on a level playing field. Like all transformations sparked by new technologies, Gaia's Web creates threats as well as opportunities.

GREEN PANOPTICON

Gaia's Web weaves together two existential threats faced by humanity in the twenty-first century. The first, a global environmental crisis, has revealed the inability of conventional political institutions to address environmental challenges like climate change and biodiversity loss. The second, a rapid technological transformation, has descended into data colonialism, surveillance capitalism, and manipulation of democracy. How could we deploy digital technology to address our most pressing environmental challenges, without falling prey to these latter risks? Is Gaia's Web to be welcomed or feared?

Let's step back and examine both sides of the debate. On the one hand, a growing movement of conservationists and scientists is creating digital solutions to environmental crises. A core assumption of their work is that better digital monitoring will be game-changing for environmental conservation, because it will address a long-standing challenge: a lack of data about Earth and its species. Advocates argue that more data will lead to better environmental protections and more efficient environmental management. On the other hand, skeptics argue that digital technology is not a magic bullet. More data, they claim, will not lead to better outcomes, particularly in cases where Big Tech pursues power and profit, and corporate control of privatized environmental data restricts access. Critics have also raised concerns over the parallels between military surveillance and eco-surveillance, identifying Digital Earth technologies as yet another risk to personal privacy. Hardcore skeptics suggest that advocating for digital technologies to solve environmental problems displays a classic error of judgment: mistaking the disease for the cure.

Amid the relentless onslaught of bad news about global environmental crises, digital technologies can seem alluring; they offer hope at a moment in which the future seems bereft, so it is tempting to believe that these technologies will rescue us. Yet entrepreneurs and the media often make exaggerated claims about the savior potential of these technologies. Worse, beyond failing to live up to the hype, these technological innovations often have negative impacts on the environment that are recognized only decades later. Some innovations that scientists initially welcomed—such as the pesticide DDT—later turned out to be catastrophic.[17]

Technologies, once released into the world, often have complex cascading effects, many of them unforeseen. Data-crunching of satellite and sensor data by AI may allow new insights into the environment, but is the high cost of digital technologies justified by positive impacts? If not, how might these funds be better spent? And what about low-income communities that do not have internet connectivity or cannot afford expensive digital devices? The Digital Earth agenda risks exacerbating the digital divide by fostering social exclusion and information asymmetries. This debate raises important questions about equity and justice. Who gains and who loses from Digital Earth?

Gaia's Web, seen in this light, is not necessarily a positive development. Yet, while recognizing the risks, my argument in this book is relatively optimistic. I remain hopeful that a progressive, digital green agenda could provide useful tools to address urgent environmental crises, allow nonhumans a degree of input and influence in environmental governance, and renew humanity's relationship with planet Earth. But there are important critiques to consider and caveats to acknowledge, given the current struggle between platform capitalism and a more democratized form of digital environmentalism. This ambivalence informs my analysis, which focuses on three questions. How could the tools of the Digital Age be mobilized to address pressing environmental challenges such as biodiversity loss and climate change? What are the potential risks and pitfalls? And, given what Sheila Jasanoff terms the "dangerous appeal of utopian technology-driven futures," how could scientists and conservationists design more inclusive, equitable alternatives?[18]

In Part I of the book, I focus on the pragmatic application of digital technology to environmental conservation issues. I explore how digital technologies offer novel ways to address climate change and biodiversity loss, advance low- or zero-waste industrial production, and enable new types of cooperative relationships in environmental regulation—perhaps even on a multispecies basis. Digital technologies allow real-time monitoring of even the world's remotest places, and AI-powered decision-making frameworks allow for the automation of environmental regulation. The organizing theme of this first part of the book is the regeneration of Earth and its ecosystems, and the advancement of conservation goals, through the practical application of digital technologies.

The theme of Part II, "Instantiating," has a double meaning: to conceptualize, in a philosophical sense, and to create new objects, in a programming sense. Here, I introduce the reader to the frontier science of biodigital engineering, ranging from synthetic biology to artificial life. In the future, our digital technologies will be increasingly fused with biological technologies via innovations like biological robots and biological computing. This will create new possibilities for environmental monitoring as well as environmental politics. This second part of the book offers a set of provocative suggestions on digital biocentrism: biodigital innovation may erode distinctions between human and nonhuman, but may also cultivate new forms of insight and understanding—even kinship—with other species.

PANDORA'S HOPE

Human ambivalence about new technology is age-old, expressed in stories and myths about our species' creative powers. In Jewish mythology, the Golem—an animate being created from mud to serve humanity—is initially servile but eventually turns against its master. In the novel *Frankenstein*, a humanoid monster created at the dawn of the Industrial Revolution struggles with existential questions, and leads its creator down a path that ultimately destroys them both. While some humans delight in their transgressive toolmaking powers, others fear—often with good reason—the consequences. Human ingenuity is both tantalizing and terrifying.

The Greek myth of Pandora reminds us that naïve hope in technology is likely to be misplaced.[19] Pandora is born in revenge, when Zeus, king of the gods, punishes Prometheus for giving fire to humans. According to Zeus's instructions, the gods molded Pandora from the mud of the Earth, an incarnation of Gaia, and presented her as a bride to Prometheus's brother Epimetheus, who foolishly accepted her despite warnings. Pandora, whose name means "all gifts," came with a dowry: a sealed box that she had been instructed never to open. But, unable to restrain her curiosity, Pandora broke the seal and opened the *pithos* (jar) that unleashed the evils, heartaches, and miseries that have plagued humanity ever since, thereby ending the Golden Age of Humanity. In horror at what she had unleashed, Pandora slammed the jar shut, trapping one spirit inside: Elpis, the *daimon* (personified spirit) of Hope. Elpis offers a parable for our dilemma about digital technology and Earth. Amid the harms sparked by human inventions, she is typically depicted as bearing an armful of flowers or gifts: a cornucopia that could be summoned by technology. Yet Elpis was also the mother of Pheme, goddess of fame and rumor, whose hunger for recognition could motivate humanity to great heights but often resulted in terrible wrongs, delusional decisions, and disastrous losses.

As an antidote to a world full of uncertainty and suffering, hope is perhaps necessary. But false hope can mislead and cloud minds. The Greek word *atê* captures this sense of self-deceiving folly. In Greek tragedies, many a doomed hero is foolishly misled by *atê* to pursue a path that leads to downfall and ruin. Yet hope—enlivening, illusory, delusional—continually resurfaces. To capture this tension, the book includes ten parables, each of which is based on an existing technology. These parables express the uncanny immanence of technology and nature, and evoke dilemmas at the confluence of digital transformation and environmental sustainability. In so doing, they remind us of the perils and promise of digital technologies that might both degrade and regenerate our relationship with Earth.

PARABLE OF TREE AND STONE

Long before dinosaurs roamed our planet, in a time called the Carboniferous, a tree was born. A seed fell to Earth, rooted in swampy soil, grew over a hundred feet tall and a hundred years long. When the tree died, its body became home to cockroaches as big as house cats, and dragonflies with wings as wide as hawks'. Bacteria fed on the rotting wood, and mosses grew. Covered by Earth's blanket, the tree's body sunk deep, compressing into coal in a slow-motion burial. Millennia later, the coal was unearthed, heated and deprived of oxygen, splintered into plastic pellets, liquified and poured into molds, and polished into small black jewels. Tree, reborn: the keys on my computer.

The stone is even older than tree.

In Precambrian time a volcano rift opened, and lava flowed from Earth's core. Cooled by rain, the lava sunk deep, compressing into stone. Millennia later, the stone was lifted from a mine shaft, crushed and bathed in caustic fluid, liquefied and poured into molds, and polished to a shine as sharp as a knife. Stone, recast: the casing for my computer, cradle for the keys.

In deep time, stones and trees are descended from stars.

Tree once drank sunlight and mixed it with air, storing energy for future generations. Stone was forged in the furnace of a long-ago star which—with the cosmic clap of a supernova—dispersed itself as stellar dust, the raw ingredient of our planet. These are the ancestors of our digital devices: Mother Tree, Father Stone, Grandmother Star, Grandfather Time.

Our computers, then, are made of stardust and tree flesh.

Their memories live on machines whose breath warms the sky.

Our digital devices are ecological, our ecologies are growing digital.

2 THE ALGORITHMIC OCEAN

Dyhia Belhabib's journey to becoming a marine scientist began with war funerals on TV. Her hometown, on the pine-forested slopes of the Atlas Mountains in northern Algeria, lies only sixty miles from the Mediterranean Sea. But a trip to the beach was dangerous. A bitter civil war raged across the mountains as she was growing up in the 1990s; the conflict was particularly brutal for Belhabib's people, the Berbers, one of the Indigenous peoples of North Africa. As she puts it: "We didn't go to the ocean much, because you could get killed on the way there."[1]

The ocean surfaced in her life in another way, on state-run television. When an important person was assassinated or a massacre occurred, broadcasters would interrupt regular programming to show a sober documentary. They frequently chose a Jacques Cousteau film, judged sufficiently dignified and neutral to commemorate the deaths. Whenever she saw the ocean on television, Belhabib would wonder who had died. "My generation thinks of tragedies when we see the ocean," she says. "I didn't grow to love it in my youth."

By the time she was ready for university, the civil war had ended. The Islamists had lost the war, but their cultural influence had grown. Engaged at thirteen to a fiancé who wanted her to become a banker, Belhabib chafed at the restrictions. Her given name, Dyhia, refers to a Berber warrior queen who successfully fought off invading Arab armies over a thousand years ago; Queen Kahina, as she is also known, remains a symbol of female empowerment, an inspiration for Berbers and for the thousands of Algerian women who took up arms in the war of independence. In a society where one in four women cannot read, Belhabib realized she didn't want to go to university only to spend her life "counting other people's money."

One day, her brother's friend visited their house. He was a student in marine sciences in the capital city, Algiers. When he described traveling out to sea, Belhabib felt a calling for an entirely unexpected path. "It was," she recalls, "a career I had never heard of, and one that challenged every stereotype of women in Algerian society." Soon after the visit, she moved to Algiers to study at the National Institute of Marine Sciences and Coastal Management, where she was one of the only women in her program. She also broke off the engagement with her fiancé, so that she could focus full-time on studies. She still vividly remembers her feelings of freedom, fear, and unreality on her first trip out to sea. While other students dove for samples, she floated on top of the water, trying to survive. "I never learned how to swim, and I still don't know how," she admits.

Belhabib graduated at the top of her class, but was repeatedly rejected when she applied to universities overseas. Her luck turned when she met Daniel Pauly, one of the world's most famous fish scientists, at a conference. Unintimidated by the fact that Pauly had just won the Volvo Prize—the environmental equivalent of a Nobel—she introduced herself and told him she wanted to study with his team. Although she did not yet speak fluent English, Pauly accepted her as a student. When she began her doctoral research, over 90 percent of the world's wild fisheries had been eradicated, and Pauly was sounding the alarm about a new, global surge in illegal fishing that was decimating marine food webs and depriving coastal communities of livelihoods. He wanted her to work on Africa, where illegal fishing had reached epidemic proportions.

Belhabib spent the next few years in West Africa. When her research uncovered the extent of illegal fishing to feed Chinese and European markets, she made the front page of the *New York Times*. "Being African myself, I was able to bring people together to openly share data in a way they never had before," she explains. It's not hard to imagine her corralling government officials: disarmingly frank and engagingly energetic, the whip-smart, hijab-wearing Belhabib stands a little over five feet tall and talks a mile a minute, with a self-deprecating laugh and a talent for gently posed, bitingly direct questions.

Her startling findings touched a nerve. Tens of thousands of boats commit fishing crimes every year, but no global repository of fishing crimes exists. A fishing vessel will often commit a crime in one jurisdiction, pay a meager fine, and sail off to another jurisdiction, thus operating with impunity. If a global database of fishing vessel criminal records could be created, Belhabib realized, there would be nowhere left to hide. She suggested the idea to a variety of international organizations, but the issue was a political hot potato; national sovereignty, they argued, prevented them from tracking international criminals. Undeterred, Belhabib decided to build the database herself. Late at night, while her infant son was sleeping, she began combing through government reports and news articles in dozens of languages (she speaks several fluently). Her database grew, word spread, and her network of informants—often government officials frustrated with international inaction on illegal fishing—began expanding. She moved to a small nonprofit and began advising Interpol and national governments. The database, christened Spyglass, grew into the world's largest registry of the criminal history of industrial fishing vessels and their corporate backers. But the registry, Belhabib knew, was useful only if the information made its way into the right hands. So in 2021 she cofounded Nautical Crime Investigation Services, a startup that uses AI and customized monitoring technology to enable more effective policing of marine crimes and criminal vessels at sea. Together with her cofounder Sogol Ghattan, who has a background in ethical AI, she named their core algorithm ADA, in homage to Ada Lovelace—the woman who wrote the world's first computer program.

Belhabib is attempting to tackle one of the most intractable problems in contemporary environmental conservation: illegal fishing.[2] Across the oceans, the difficulty of tracking ships creates ideal cover for some of the world's largest environmental crimes. After the end of World War II, the world's fishing fleets rapidly industrialized. Wartime technologies that had been developed for detecting underwater submarines were repurposed for spotting fish. The size of nets grew exponentially, and offshore factory ships were outfitted so they could spend months at sea, extending the reach of industrial fishing into the furthest reaches of the ocean. As the world's population grew, fish protein became an increasingly important source of food. But warning signs soon appeared: crashes in key fish populations, an alarming trend of "fishing down marine food webs," and a series of cascading impacts that rapidly depleted marine ecosystems.

In the wake of depleting stocks, fishers should have responded by reducing their take. Instead, they redoubled their efforts. After the world's leading fishing nations—China and Europe are the largest markets—overfished their own waters, they began exporting industrial overfishing to the global oceans. China's offshore fishing fleet of several hundred thousand vessels, which received nearly $8 billion in government subsidies in 2018, is now the largest in the world.[3]

Governments of wealthier nations subsidized massive fleets of corporate-backed vessels to fish the high seas, using bottom trawling and drift nets stretching for dozens of miles, killing everything in their path. Artisanal fishers were squeezed out, and as fish stocks collapsed, rising food insecurity generated protests and political unrest. In West Africa, for example, fishing boats from the world's wealthiest nations have depleted local fisheries to such an extent that waves of migrants—faced with food insecurity and uncertain futures—have begun fleeing their homes in a desperate, risky attempt to reach European outposts such as the Spanish Canary Islands; thousands of migrants have died at sea. The smaller fishing fleet, meanwhile, has struggled to remain solvent; impoverished fishers are increasingly vulnerable targets for criminal organizations seeking mules for hire to transport drugs, or boats to serve as cover operations for human trafficking.[4]

Over 90 percent of the world's fish stocks are now fished to capacity or overfished.[5] Despite this, scientists' calls for reduced fishing have largely fallen on deaf ears. Conventional attempts to manage fisheries are stymied by the limits of logbooks and onboard human observers, and local electronic monitoring systems. Fishing boats that exceed quotas or fish in off-limits areas are rarely caught, operating with impunity in front of local fishermen's eyes; and even if caught, they are even more rarely punished.

MARINE PANOPTICON

The world's oceans are experiencing an onslaught: as fish have become scarcer, illegal fishing has surged. Rather than merely document the decline of fish stock, Belhabib decided to do something about it. Her solution: to combine ADA, her AI-powered database of marine crimes, with data that tracks vessel movements in real time. She began by tracking signals from the marine traffic transponders carried by oceangoing ships—also known as automatic information systems (AIS). AIS signals are detected by land transceivers or satellites and used to track and monitor individual vessel movements around the world.[6] AIS signals are also detected by other ships in the vicinity, reducing the potential for ship collisions. Belhabib and her team then built an AI-powered risk assessment tool called GRACE (in honor of the pioneering coder Grace Hopper), which predicts risks of environmental crimes at sea. When combined with vessel detection devices such as AIS, GRACE provides real-time information on the likelihood of a particular ship committing environmental crimes, which can be used by enforcement agencies to catch the criminals in the act. Belhabib's database means that criminal vessels—which often engage in multiple forms of crime, including human trafficking and drug smuggling, as well as illegal fishing—now find it much harder to hide.

The high seas are one of the world's last global commons, largely unregulated. The UN Convention on the Law of the Sea provides little protection for the high seas, two-thirds of the ocean's surface. The adoption of a new United Nations treaty on the high seas in 2023 will create more protection, but this will require years to be implemented. Even within 200 nautical miles of the coast, where national authorities have legal jurisdiction, most struggle to monitor the oceans beyond the areas a few miles from the coast. And beyond the 200-mile limit, no one effectively governs the open ocean.

So Belhabib hands her data on human rights and labor abuses over to Global Fishing Watch, a not-for-profit organization that collaborates with the national Coast Guards and Interpol to target vessels suspected of illegal fishing for boarding, apprehend rogue fishing vessels, and police the boundaries of marine parks. The observatory visualizes, tracks, and shares data about global fishing activity in near real time and for free; launched at the 2016 US State Department's "Our Ocean" conference in Washington, it is backed by some of the world's largest foundations. Its partners include Google (which provides tools for processing big data), the marine conservation organization Oceana, and SkyTruth—a not-for-profit that uses satellite imagery to advance environmental protection.

Global Fishing Watch uses satellite data on boat location, combined with Belhabib's data on criminal activity, to train artificial intelligence algorithms to identify vessel types, fishing activity patterns, and even specific gear types (tasks that would require human fisheries experts hundreds of years to complete).[7] The tracking system pinpoints each individual fishing vessel with laser-like accuracy, predicts whether it is actually fishing, and even identifies what type of fishing is underway. Their reports have revealed that half of the global ocean is actively fished, much of it covertly. Fred Abrahams, a researcher with Human Rights Watch, explains that this approach is just one example of a new generation of conservation technology that could act as a check on anyone engaged in resource exploitation. His team at Human Rights Watch uses satellite imagery to track everything from illegal mining to undercover logging operations. As Abrahams says: "This is why we are so committed to these technologies . . . they make it much harder to hide large-scale abuses."[8] Abrahams, like other advocates, is confident that the glitches—for example, AIS tags are not yet carried by all fishing vessels globally, poor reception makes coverage in some regions challenging, and some boats turn off the AIS when they want to go into stealth mode—will eventually be solved. Researchers have recently figured out, for example, how to use satellites to triangulate the position of fishing boats in stealth mode—enabling tracking of so-called dark fleets. These results can inform a new era of independent oversight of illegal fishing and transboundary fisheries. Meanwhile, researchers are developing other applications for AIS data, including assessments of the contribution of ship exhaust emissions to global air pollution, the exposure of marine species to shipping noise, and the extent of forced labor—often hidden, and linked to human trafficking—on the world's fishing fleets.

It's a herculean task for one organization to police the world's oceans. And Global Fishing Watch's data is mostly retroactive; by the time the data is analyzed and the authorities have arrived, fishing vessels have often left the scene. What is still lacking is a method for marine criminals to be more effectively tracked in real time, and apprehended locally. This is where Belhabib's next venture comes in. She is now working with local governments in Africa—where much illegal fishing is concentrated—to provide them with trackers and AI-powered technologies to catch illegal fishing and other maritime crimes in the act. As she notes: "When you ask the Guinean Navy how much of their territorial waters they can actually monitor, it's only a fraction of a vast area. They simply don't have the resources." Belhabib's system pinpoints vessels that may be committing infractions, and assesses the risk live on screen. This allows the Coast Guard and other agencies such as Interpol to more easily find illegal fishers, while reducing the costs of deployment, monitoring, and interdiction.

She cautions, however, about the use of similar digital technologies to track illegal migrants. The European Union, for example, has strengthened its "digital frontier" through satellite monitoring, unmanned drones, and remotely piloted aircraft, in some cases relying on private security and defense companies to undertake data analytics and tracking. But these technologies are often focused on surveillance rather than search and rescue of migrants stranded at sea.[9] As Belhabib relates: "Recently I spoke with the Spanish Navy and they told me they watched over 100 people die when a boat full of migrants capsized and they could only save a few people. They told me, 'We take their fish away, they risk their lives to have a better and decent life.' It's heartbreaking and avoidable." In Belhabib's view, Digital Earth technologies should prioritize ecological and humanitarian goals, rather than surveillance and profit.

To recap my argument thus far: Digital Earth technologies enable more rapid detection and, in some cases, prediction of marine crimes. Digital monitoring, combined with artificial intelligence, allows precise analysis of fishing vessel locations and movements at a global scale. Although this does not guarantee enforcement, it could enable more efficient policing of the world's oceans. The use of digital technologies enables conservationists to tackle two common flaws that lead to failures in environmental enforcement. First: data is scarce; if available, there is often a time lag, geographical gaps, or data biases. This makes evidence-gathering difficult or impossible. Second, enforcement often comes too late. Environmental criminals can be prosecuted, but legal victories are uncertain, and happen after the damage has been done. These shortcomings of contemporary environmental governance—sparse data, unenforceable regulations, and patchy, sporadic enforcement that punishes but fails to prevent environmental harm—can be overcome by digital monitoring, which mobilizes abundant data in real time to gather systematic evidence and enable timely enforcement.

These techniques appear to be achieving some success.[10] In Ghana, for example, there has been a long-standing conflict between industrial fishing boats and small-scale, artisanal fishers using canoes and small boats to fish near the shore. Satellite data has helped the government's Fisheries Enforcement Unit track and reduce the incursions of larger fishing boats into near-shore waters.[11] In Indonesia, the world's largest archipelago country with the second-longest coastline in the world, the government has entered into an agreement with Global Fishing Watch data to monitor fisheries and share the data about vessels' movements publicly online, a major step forward in transparency in fisheries enforcement. The Indonesian partnership is an example of the longer-term aim of Global Fishing Watch: to share its geospatial datasets and online mapping platform with governments around the world.

Despite these recent gains to combat illegal fishing, digital tech is also exacerbating the underlying problem, as fishers themselves have started taking advantage of digital strategies. One example is the growing use of fish aggregating devices, which use acoustic technology, combined with satellite-linked global positioning systems, to better spot schools of fish. Fishers can effectively assess location, biomass, and even species, allowing them to aggregate and fish more efficiently. Digitization is ratcheting up the already intensely competitive fishing industry and accelerating the overfishing of endangered species.[12]

Even if conservationists can win this digital arms race, there is a more fundamental problem: the underlying structural drivers of overfishing—consumer demand, particularly in Asia and Europe, and a lack of adequate governance for the high seas—are not solvable by digital technologies alone. Governance reform and digital innovation must work in tandem. For example, in the absence of government regulation, digital monitoring of fishing on the open ocean would be unlikely to scale up. But the adoption of the new UN treaty on the high seas in 2023 included a significant commitment to creating new Marine Protected Areas, aligned with Global Biodiversity Convention's commitment to protect 30 percent of the Earth's land and oceans by 2030. These new developments create an impetus for digital monitoring; and, in turn, digital monitoring will increase the likelihood that Marine Protected Areas will be effective at protecting fish populations. This illustrates two key points about environmental governance in the twenty-first century: the interplay between digital and governance innovation, and the fact that planetary governance of the environment is possible only with planetary-scale computation.

MISSION TO PLANET EARTH

Conservationists such as Belhabib are deploying cutting-edge digital technologies that have been technically possible to develop only in the past few years. Yet the ideas behind these technologies date back more than fifty years. In December 1968, the Apollo 8 mission launched three astronauts into space, embarking on a journey farther than any human had ever traveled. The spacecraft orbited the moon ten times without landing; at the farthest reach of each orbit, the Earth disappeared from view, and all radio contact was lost. Until Earth rose again on the horizon, only the mysterious far side of the moon—never before observed by humans—was in view, its pitted surface empty of life. As Earth reappeared, small and blue in the distance, an astronaut took a photograph of our planet floating in the immensity of space. *Earthrise* became an iconic image, inspiring a new appreciation for the beauty and fragility of life.[13] Appeals by returning astronauts to protect nature, and their description of mystical experiences induced by observing Earth from afar, captured the imagination of a generation.

Embedded within the spacecraft was the first computer in the world to rely on silicon-based integrated circuits. The astronauts, trained pilots, did not actually fly the spaceship; rather, they used control sticks to command the Apollo Guidance Computer, which fired thrusters to move the lunar module. Every move was mediated by software, written by young coders, which connected the onboard computer to the control center back on Earth thousands of miles away. What was referred to as "manned" spaceflight was, in fact, a collaborative human-machine endeavor. Humanity's newfound appreciation of our planet, adrift in the immensity of space, was made possible by digital technology.[14]

The *Earthrise* image circulated at a time of political upheaval. Only a few years prior, the Cuban missile crisis intensified collective dread about the ever-present threat of nuclear war between the Soviet Union and the United States. In years following the mission, environmental issues—the exhaustion of natural resources, chemical pollution, and destruction of biodiversity—moved to the top of public awareness, culminating in the first Earth Day in 1970 and the world's first Earth Summit in 1972.

This was also a time of upheaval in planetary science. Just prior to the Apollo 8 mission, a revolutionary idea that described the Earth's subsurface was gaining ground: plate tectonic theory, which describes the Earth's surface as a set of plates that move relative to one another.[15] At the same time, scientists were developing a new framework for describing biogeochemical models of Earth, in which elements such as carbon circulate between the atmosphere and the Earth's crust or lithosphere. Earth system science, geobiology, astrobiology, and other scientific fields were created to explore how our planet had evolved, and how life both adapted to environmental change and simultaneously acted as a geological force. The interplay between living and nonliving elements of the Earth, understood as a planetary system, became the object of intense scientific study. Marshall McLuhan, commenting on the Soviets' Sputnik launch in 1957, predicted that the Space Race would render Earth's observations computable.[16] The beginnings of digital observation of Earth were rooted in a time when the Space Race, and the potential discovery of life on other planets, inspired both scientific inquiry and the public imagination.

As discussed in the previous chapter, the Gaia metaphor, developed by Lynn Margulis and James Lovelock, popularized this concept for mass audiences. Both Margulis and Lovelock worked for a time at NASA, where they met while collaborating with scientists searching for life on other planets.[17] The successful moon landing had sparked political interest in colonizing the solar system, sparking intense geopolitical competition and raising speculation about the existence of life elsewhere in the universe. NASA's SETI (Search for Extra-Terrestrial Intelligence) initiative—rivaled by the Soviet space program and its parallel CETI program—pursued a joint scientific and geopolitical agenda: conquering space.[18] The Space Race included an aggressive program of telescope development and satellite launches.

Although many satellites were for military purposes, civilian applications were also of interest, and the first Earth observation satellites were launched in the 1970s. To address the challenge of observing other planets from afar to detect signs of life, scientists turned to the easiest available experimental object at hand: Earth. Margulis developed the idea that some atmospheric gases could be biological in origin, originally produced by bacteria—a controversial view at the time. When she reached out to Lovelock, who was engaged in the study of atmospheric chemistry at a planetary scale, the concept of Gaia—a self-regulating system of biotic and abiotic elements, which creates and sustains atmospheric conditions supportive of life on Earth as we know it—was born. The composition of a planet's atmosphere, they hypothesized, could be interpreted from afar to assess signs of life.

Margulis and Lovelock soon turned their attention to our planet itself. In their decades-long collaboration on Gaia, they reframed Earth as the largest of ecosystems, in which biota share and exchange water, nutrients, and energy with one another and their environment. Their work evolved in tandem with NASA's program of planetary observations from space and digital representations of Earth. Lovelock created a computer simulation called Daisyworld, which demonstrated the central hypothesis of Gaia theory: living organisms can affect the entire climate system.[19] The Daisyworld model contained only two species—black daisies and white daisies—and a few coupled equations. Yet it signified a much bigger idea: with enough data, one could model the entire Earth as a single system. This concept, it turns out, was prescient, a harbinger of today's massive efforts at planetary-scale computation.

THE ECO-TECH STACK

The descendants of Daisyworld are today's contemporary climate models. Run on supercomputers and made up of millions of lines of code, these complex computer programs simulate the physical and biochemical processes of the Earth's climate system, modeling temperature, oceans and sea ice, cloud formation and rainfall, sunlight and vegetation. Less than a century ago, atmospheric scientists trudged up remote volcanoes or were towed in small dinghies behind boats in order to collect atmospheric measurements. Today, these measurements are instantaneously available, embedded in computational systems that are increasingly accurate at predicting the complex ebbs and flows of our climate, and its relationship to other parts of the Earth system. Climate models are one important element of the digital networks used to monitor Earth, its ecosystems, and its inhabitants for environmental management and conservation purposes.

Digital environmental monitoring thus extends to ecological systems what engineers conventionally refer to as a tech stack—the combination of technologies used to run a system or deliver a product. Benjamin Bratton refers to this extended stack as a planetary scale set of infrastructures: a global network of networks, composed of the hardware and software that support cloud servers, smart grids, robotics, smartphones, and ubiquitous computing.[20] Although much attention has been paid to the evolution of the tech stack, one important aspect is often overlooked: the increasing integration of Earth's ecosystems with digital technology. To understand this point, let's look at how the eco-tech stack is actually built. On the one hand, it includes components familiar to engineers: interconnected systems of satellites and sensors that are instrumented, interconnected, and intelligent: digital instrumentation, including sensors and satellites; interconnected monitoring systems and data architectures; artificial intelligence; and data architectures and decision-support systems that enable information on environmental issues to be conveyed in an automated fashion, and often in real time (figure 2.1). But it also includes components that would be less familiar to most engineers, including living beings—humans and nonhumans who gather data and are connected to the network. Before I expand on this latter point, let's look more closely at the former.

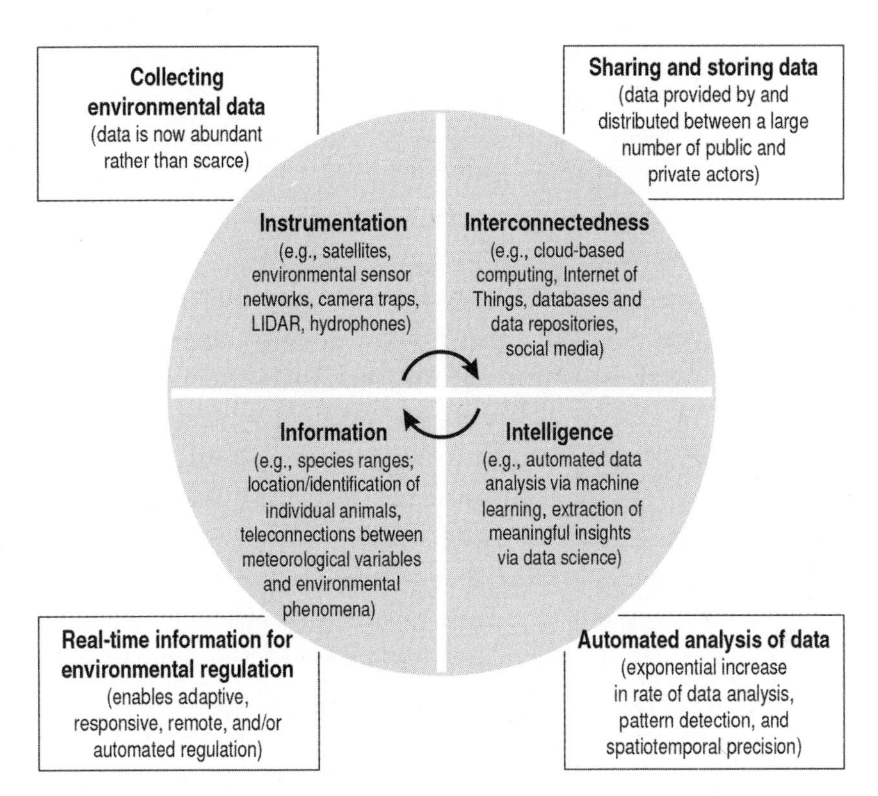

Figure 2.1
Digital Earth technologies. Source: Author.

When Congress canceled SETI after the end of the Cold War, Earth observations became the primary focus of the majority of the thousands of satellites now orbiting the planet, the most extensive of which was NASA's "Mission to Earth" system of satellites for environmental observations—the Earth Observing System Data and Information System (EOSDIS). EOSDIS was the first space-based program that sought to assess humanity's impact on Earth's environment, and also one of the first to make global environmental data available to the world.[21] The underlying technology was cumbersome: the earliest American satellites used photographic film, which was jettisoned in a reentry capsule that had to be caught mid-air by a US Air Force plane using a purpose-built retractable mechanical arm nicknamed "The Claw."[22] Early on, the data generated by satellites was laborious to analyze, access was limited, and satellites were enormously expensive to build and launch. But satellite technology rapidly improved, enabling the development of remote sensing technology, which used satellite imagery to monitor everything from weather, land-use types, forest fires, desertification, polar ice, sea levels, water circulation in Earth's atmosphere, coral reef bleaching, and ocean conditions.

Over the past few decades, a revolution has occurred in satellite technology. Satellites are cheaper, in part because they are smaller: nano-satellites are now just slightly larger than a big knapsack, while the smallest "satlets" are smaller than a standard piece of paper and no more than a few inches thick.[23] Whereas earlier generations of satellites cost millions of dollars and were loaded with expensive instrumentation that quickly became outdated (sometimes even before launch), today's modular satellites are more flexible, more easily updated, and orders of magnitude cheaper. In the past, most satellites were owned by national governments. The public, and even scientists, often had limited access to data. Satellites were stand-alone observing devices, and coverage was often patchy. Today's satellites, many of which are privately owned, can connect to one other in networks. Data from these networks can be combined with data from other sources. A constellation of data sources, gathered and controlled by both private and state actors, has replaced the more tightly controlled nation-state satellite programs of the past. The tsunami of digital data has sparked innovation in environmental and geospatial data analytics, including private companies offering geospatial data refineries, and open-source projects offering data management tools, extending access to Digital Earth data much more widely than in the past. For example, Open Data Cube technology enables continental-scale land cover mapping with geospatial data; it has been adopted by more than fifty countries and is enabling a nonprofit, Digital Earth Africa, to assemble and share environmental data for the entire continent.

Within Digital Earth networks, data derived from remote monitoring is often integrated with data from sensors in the water or on land, or carried by drones. Environmental sensors are usually electronic, but as explored in the final chapters of this book, environmental monitoring also uses biosensors and biohybrid robots that are part organism, part machine. These sensors generate enormous amounts of fine-grained data about Earth and its inhabitants, ranging from biophysical data to biological data. The size and price of these sensors has dropped dramatically. When combined, these Earth imaging systems can provide rapid sampling intervals and global coverage of the planet's surface, measuring everything from greenhouse gases and polar ice to water circulation and ocean conditions to the Earth's magnetic field and solar radiation, and even soil contamination.[24]

The eco-tech stack also involves humans and nonhumans participating as sensing devices. The worldwide growth of Digital Earth technologies potentially converts every citizen into an environmental sensing device.[25] If you enroll in an environmental monitoring program, the app on your phone will use its sensors to collect data on environmental conditions like temperature or subtle seismic vibrations. Even if you do not enroll in such a program, environmental data that you unwittingly gather on a daily basis might be distributed or sold without your knowledge.[26] Nonhumans, too, are deployed as sensors.[27] For example, the ICARUS network launched on the International Space Station in 2018 relies on a new generation of satellite monitoring transmitter tags that can weigh less than a gram; light enough to attach to a honeybee, they are still powerful enough to send a signal to a satellite. Some tags store hundreds of megabytes of data (enough to last most animal lifetimes) and don't even need to be retrieved; the data can be downloaded with a computer or smartphone. ICARUS uses these tags to track animals around the globe, in near real time. Its first subject: a Eurasian blackbird, tagged in Belarus, which eventually flew over 1,000 miles to Albania, tracked by the International Space Station orbiting nearly 250 miles overhead.

The founder of ICARUS, German scientist Martin Wikelski, envisions an Internet of Animals that flips the traditional model of satellite-based Earth observations on its head: rather than relying on humans to interpret images of Earth's surface captured by satellites, animals acting as sensors will provide real-time information on the quality and health of ecosystems, as well as the behavior and physiology of animals themselves. Similar to a smartphone traffic app, ICARUS can provide information, in real time and in aggregate, about animal movements that reveals much about weather, climate, habitat integrity, ecosystem health, and even disease. Compared with manual tagging methods (which recover, in the case of birds, less than 1 percent of tags), ICARUS begins a new era in animal sensing. Eventually, Wikelski envisions a network of 100,000 animals (ICARUS has already tagged birds and bats, goats, rhinos, and tortoises) that act as animal sentinels, warning of environmental change, and even potential animal reservoirs of infectious diseases. Other projects are using bird-borne samplers for monitoring the atmosphere, whale-borne sensors for monitoring the ecological health of the oceans, and albatrosses for monitoring illegal fishing.[28] Once sufficient big data studies of animal movements have been conducted, researchers intend to build real-time predictive models that integrate animal movement patterns, habitat preferences, and animal memories into movement forecasts. This would enable the use of mobile protected areas that follow animals on the move, monitoring and protecting them as they migrate. As scientists developing the ICARUS program explain: "We suggest that a golden age of animal tracking science has begun and that the upcoming years will be a time of unprecedented exciting discoveries that . . . do more than inform us about particular species of animals, but [also] allow the animals to teach us about the world."[29]

ICARUS will collect a much broader array of data than previous initiatives, logging not only an animal's location but also its physiology and environment. "It's a new era of discovery," says Walter Jetz, a professor of ecology and evolutionary biology at Yale University, whose center is working with the project. "We will discover new migration paths, habitat requirements, things about species behavior that we didn't even think about. That discovery will bring about all sorts of new questions." Founder Martin Wikelski imagines a future where ordinary people can follow their favorite fish, bear, or bird in real time, enabled by the space station's tracking of the creature overhead. Similar tracking systems have been set up for whales and seals, birds and honeybees. Many of these Digital Earth animal tracking initiatives overlap with the emerging subdisciplines of ecological informatics and "movement ecology," which use miniaturized radio and acoustic transmitters, cellular and satellite networks, global positioning systems, and light-level geolocators to track organisms for longer periods of time, in ever-remoter areas.

Environmental sensing technology is rendering our planet itself computable. In doing so, computation has been rendered environmental. The tech stack is no longer purely technological, but also ecological. And the feedback loop continues: as we will later discuss, Digital Earth technologies use tremendous amounts of electricity and resources and also generate tremendous amounts of waste; they now influence the biogeochemical cycles without which life on Earth would cease to exist. Gaia's Web is digital and biological, geological and ecological.

PLANETARY COMPUTATION, PLANETARY GOVERNANCE

What are the implications of Gaia's Web for environmental governance? Digital Earth technologies improve environmental governance in two ways. First, using relatively low-cost, digitally automated monitoring technologies, environmental conditions can be accurately mapped. This makes it easier to detect environmental impacts and crimes, even in hard-to-reach areas. Geographical ubiquity is a hallmark of digitally enabled twenty-first-century environmental governance. For example, satellites are now being used to identify greenhouse gas emissions like methane in real time; NGOs are publishing "name and shame" lists of the world's biggest climate polluters.

Second, Digital Earth technologies are enabling environmental governance to operate at much higher speed. In the past, environmental regulators were often slow to respond; eventually, the polluter of a river might be prosecuted, but the fish would already have died. Digital Earth technologies enable continuous monitoring, and detection and potentially enforcement, when the environmental harm is starting to occur (table 2.1). This is akin to the real-time, responsive digital technologies used in Smart Cities, but expanded to include a much wider range of ecosystems. By increasing the capacity to detect polluters and poachers in the act, regulators can enhance deterrence. The everywhere-all-the-time capability of Digital Earth monitoring, driven by rapid decline in the cost of monitoring technologies, has increased scientists' ability to conduct real-time assessment of environmental changes. This development is significant, as it implies the time-space

Table 2.1

Real-time environmental monitoring

Type of regulation	Issue	Application
Real-time response to poaching and illegal resource extraction	Illegal logging	Invisible Tracck: Tracking device on tree trunks that alerts responders to illegal logging[a]
	Poaching	Instant Detect 2.0: Camera trap system that uses satellite technology to stop poaching by sending data instantly and alerting responders; Air Shepherd: Uses SPOT (unmanned aerial vehicles with thermal infrared cameras) that can detect and intercept poachers[b]
	Smuggling	InvestEGGator: Fake 3D-printed turtle eggs with a GPS sensor used to track smugglers and alert customs authorities in order to make arrests[c]
Real-time disaster detection/warning	Floods	RiverTrack: A system of sensors placed along waterways that continuously informs local residents about the water level, sending alerts when the water reaches a level of concern[d]
	Ice thinning	SmartICE: A sensor system designed to detect sea ice thickness and support the safe travel of Inuit people over sea ice, in the face of changing climate and thinning ice (includes SmartBUOY and SmartQAMUTIK)[e]
	Forest fires	FireALERT MK I: A system of sensors that detects the spectral patterns of fire and sends real-time alerts while continuously tracking the status and location of the fire[f]
Real-time environmental monitoring	Air quality	Green Horizons: Creates maps displaying the concentration and dispersion of pollutants across Beijing[g]
	Freshwater	WaterWatch: Maps of real-time water data that include current streamflow, drought, flood, past flow, and runoff[h]
	Endangered species identification	WildBook: Combines computer vision system and crowdsourcing to enable identification and accurate censuses of endangered species[i]
	Animal migration	ICARUS satellite: Global animal observation system via satellite, supported by Movebank (user data center for animal tracking datasets worldwide); enables users to watch animal movements in real time[j]
	Climate change	Global Tagging of Pelagic Predators: Geotagging of ocean animals to observe climate change and effects on habitat[k]

Table 2.1 (continued)

Type of regulation	Issue	Application
Real-time resource/ extractive industry monitoring	Fugitive methane emissions	MethaneSAT: Real-time tracking and managing of methane leaks[l]
	Agriculture	SenseHub: Technology that gathers data about livestock from a mounted tag sensor, which helps farmers respond to livestock needs in near real time[m]
	Mining	SENSEI: Real-time monitoring of groundwater in mining operations; Internet of Things (IoT) applications that enable real-time management, thereby decreasing environmental impact and increasing resource efficiency[n]

a. Elga Reyes, "Innovative Monitoring Systems Stop Illegal Logging in Real Time," *Eco-Business*, July 19, 2013, https://www.eco-business.com/news/innovative-monitoring-systems-stop-illegal-logging-real-time/.

b. "Instant Detect," Zoological Society of London, accessed May 8, 2023, https://www.zsl.org/conservation/how-we-work/conservation-technology/instant-detect; "Machine Learning for Wildlife Conservation with UAVs," Harvard University TEAMCORE, accessed May 9, 2023, https://teamcore.seas.harvard.edu/machine-learning-wildlife-conservation-uavs.

c. Sue Palminteri, "USAID Wildlife Crime Tech Challenge Awards Acceleration Prizes for Rapid Tech Developments," *Mongabay*, October 2, 2017, https://news.mongabay.com/wildtech/2017/10/usaid-wildlife-crime-tech-challenge-awards-three-acceleration-prizes/.

d. Scottish Environment Protection Agency, "Smart Tech Solution Helps Scots Fight Flooding," *Scottish Environment Protection Agency*, June 22, 2018, accessed May 9, 2023, https://media.sepa.org.uk/media-releases/2018/smart-tech-solution-helps-scots-fight-flooding/.

e. "Our Smart Technology," SmartICE, accessed May 9, 2023, https://smartice.org/our-smart-technology/.

f. "Wildfire Detection," *Vigilys*, accessed May 8, 2023, https://vigilys.com/technology/firealert/.

g. IBM, "Green Horizons," IBM Media Center, video, 4:44, October 11, 2018, https://mediacenter.ibm.com/media/Green+Horizons/1_f9ftqtn1.

h. "About WaterWatch," US Geological Survey, accessed May 9, 2023, https://waterwatch.usgs.gov/index.php?id=ww_about.

i. Tanya Y. Berger-Wolf, Daniel I. Rubenstein, Charles V. Stewart, Jason A. Holmberg, Jason Parham, Sreejith Menon, Jonathan Crall, Jon Van Oast, Emre Kiciman, and Lucas Joppa, "Wildbook: Crowdsourcing, Computer Vision, and Data Science for Conservation," *arXiv:1710.08880* (2017).

j. "About Icarus," ICARUS, accessed May 9, 2023, https://www.icarus.mpg.de/28056/about-icarus; "Animals on the Air," ICARUS, accessed May 9, 2023, https://www.icarus.mpg.de/28874/sensor-animals-tracking.

k. Elliot L. Hazen, Salvador Jorgensen, Ryan R. Rykaczewski, Steven J. Bograd, David G. Foley, Ian D. Jonsen, Scott A. Shaffer, et al., "Predicted Habitat Shifts of Pacific Top Predators in a Changing Climate," *Nature Climate Change* 3, no. 3 (2013): 234–238.

l. "The Satellite," MethaneSAT, accessed May 9, 2023, https://www.methanesat.org/satellite/.

m. "Solutions," Allflex Livestock Intelligence, accessed June 10, 2020, https://www.allflex.global/livestock-monitoring/; "SenseHub™ Cow Calf: Helping You Improve Conception Rates," Merck, accessed May 9, 2023, https://merck.sensehub.global/cowcalf/.

n. David Simpson, "Real-Time Groundwater Monitoring," accessed May 9, 2023, https://www.csiro.au/en/work-with-us/industries/mining-resources/resourceful-magazine/issue-19/real-time-groundwater-monitoring.

compression of environmental data. In turn, this enables real-time regulatory responses. A gas well leaking methane in Pennsylvania can be detected (and named and shamed) by a flock of nano-satellites launched by ex-NASA engineers based in San Francisco. A coffee drinker in a supermarket in Hyderabad can scan a QR code on a bag of coffee, learn about the Costa Rican farmer who produced the coffee beans, and even donate to the farmers.

Proponents of Digital Earth conservation argue that sufficient data, if accessible in an equitable and transparent manner, could create powerful new ways to govern the environment: detecting and monitoring, and even predicting and preventing environmental hazards; reducing pollution, notably climate change emissions; and tracking environmental degradation and biodiversity loss in real time. Yet although these applications offer potential to reduce environmental harm, critics warn of pitfalls. They question, for instance, whether more data is necessarily better. The continuous monitoring of the environment, and the integration of hundreds or thousands of variables, creates a data deluge; environmental data becomes hyper-abundant rather than scarce. Collecting sufficient data and successfully integrating it is technically difficult, due to limitations of digital devices—resolution, battery life, coverage—and artificial intelligence algorithms, which require substantial training and manual labeling of datasets, are extremely costly and energy-intensive to use, and are far from perfectly reliable. Further difficulties arise due to the different requirements posed by monitoring different aspects of the Earth system; for example, migratory animals require very different digital monitoring regimes than plants. A further difficulty arises when integrating data over the varying timescales over which environmental change occurs, from seconds to millennia.

However, if these challenges are addressed, the eco-tech stack creates preconditions for potentially significant shifts in environmental governance: real-time, predictive, and hence preventative, automated, spatially ubiquitous, and collaborative regulation. Yet even if technologically feasible, such an improved monitoring system is by no means guaranteed to lead to better environmental outcomes. As explored in the chapters which follow, some Digital Earth innovations are reminiscent of Bentham's panopticon: a ubiquitous regime of surveillance that could easily be misused as a mechanism of social control. Citizens interacting with Digital Earth technologies submit to surveillance, providing data while having their online actions indexed. This raises the potential for inadvertent (or deliberate) "bycatch" of humans participating in environmental monitoring, creating potential privacy risks. The expansion of surveillance is inherently double-edged, as Foucault's critique of the panopticon reminds us. Digital Earth is a biopolitical technology, which seeks to administer and control life through ubiquitous systems of control of living beings. If left unchecked, it is likely that many digital technologies discussed in this book could advance this paradigm of biopolitical control, aligned with surveillance capitalism.[30]

Moreover, Digital Earth governance may exacerbate the problem of algorithmic bias, as it shifts environmental regulation to automated decision-making systems. Such biases might express structural inequalities of race, gender, or species membership. Specific individuals and communities might be unfairly targeted, or summarily excluded, by automated tracking systems. Scholars have also voiced concerns that citizen sensing is not, a priori, collaborative and democratic. Do data flows feed into predetermined interests of experts, scientists, and corporations, or do they enable citizens to provide meaningful input? Crowdsourcing may ironically entrench the expert hierarchies it purports to challenge. These concerns are further underscored by the increasing interest of Big Tech in digital environmental monitoring.

AS EARTH COMES ONLINE, WHO IS IN CONTROL?

In 2017, Microsoft appointed its first Chief Environmental Scientist, Lucas Joppa, with a mandate to develop "Earth algorithms." The company invested over $50 million in its new AI for Earth program. As Joppa wrote in 2019 in *Scientific American*:

> The epitome of the innovation we need is best understood as a planetary computer. A planetary computer will borrow from the approach of today's internet search engines, and extend beyond them in the form of a geospatial decision engine that supports queries about the environmental status of the planet, programmed with algorithms to optimize its health. Think of this less as a giant computer in a stark white room and more of an approach to

computing that is planetary in scale and allows us to query every aspect of environmental and nature-based solutions available in real time. We currently lack the data, compute power and scalability to do so. Only when we have a massive amount of planetary data and compute at a similar scale can we begin to answer one of the most complex questions ever posed—how do we manage the earth's natural resources equitably and sustainably to ensure a prosperous and climate-stable future?[31]

The following year, Microsoft president Brad Smith formally launched the Planetary Computer, a computing system that aggregates global environmental data into a searchable dataset, noting, "It should be as easy for anyone in the world to search the state of the planet as it is to search the internet for driving directions or dining options."[32] Whether global environmental change is as easy to interpret as a restaurant menu or street map is debatable. But the broader point is clear: by translating the planet into digital data, Microsoft—and many other major tech companies—is incorporating sustainability into its digital innovation agenda. As Amy Luers, Microsoft's Global Director for Sustainability Science, wrote soon after: "Achieving global sustainability goals will require developing planetary intelligence. Planetary intelligence for sustainability encompasses the following: collecting information at a planetary scale and using that information to understand the Earth as one interconnected social-biophysical system, assessing and predicting implications of changes for people and nature, providing transparency and accountability in environmental management, and shaping markets and policies to achieve sustainable outcomes."[33]

Many of the characteristics of Digital Earth governance—real-time, predictive, dynamic, data-rich—are outlined in Microsoft's agenda. As mentioned in chapter 1, nearly all the major tech companies have similar initiatives: IBM's Smarter Planet, Nokia's SensorPlanet, HP Labs' Central Nervous System for the Earth (CeNSE), NASA's Earth Observing System Data and Information System (EOSDIS), Cisco/NASA's Planetary Skin Institute, and China's Digital Earth initiative. Under its Google Sustainability umbrella, Google runs Google Earth, Google Bioacoustics, Google Ecology, and the Google Climate Change Summit. Critics have raised concerns over the parallels between surveillance and eco-surveillance, and corporate control of environmental data, and Big Tech's engagement is one reason that Digital Earth governance has been subject to growing critiques. Who controls environmental data? Are Big Tech companies creating an environmental version of what Shoshana Zuboff calls surveillance capitalism?[34] What will ensure that these innovations are used as relatively benign tools rather than deployed as dangerous weapons?

Discussions of digital technology often oppose two extreme views of utopian or dystopian futures. Consider the cover story of *National Geographic* in April 2020, in recognition of the fiftieth anniversary of Earth Day, on April 22. The front cover offers a grim picture of a dying planet in which humanity loses the battle against climate change. But the back cover offers a strikingly contradictory scenario: global salvation through technological innovation, averting climate catastrophe. For the same Earth Day celebration, the cover of *Wired*—Silicon Valley's beloved monthly magazine—showcased a photo of Earth in the rearview mirror of a departing spaceship, with the tagline: "What will keep us human when we leave the Earth behind?" Rather than saving the planet, which was pictured on the cover as a delectable scoop of rapidly melting, multicolored ice cream, *Wired* urged its readers to imagine "the care and feeding of interplanetary civilization," with practical advice for escapees of our ravaged planet: cafeteria menus on future spaceships, satellite monitoring of climate change–fueled superstorms, floating farms designed to survive apocalypse.

These stories express divergent worldviews. The view of Dominion, also referred to as eco-modernism, asserts that humans are endowed with special status—whether religious, spiritual, or intellectual—to govern Earth-as-Machine; humans should deploy technology to exploit, manipulate, and reshape the Earth for human benefit. Communion, a romantic view favored by environmentalists, holds that Earth-as-Eden has been disrupted by humans, but that humans can restore balance and harmony by restraining our destructive tendencies and learning to love the Earth. Escapism, favored by technologists, frames Earth-as-Throwaway-Cradle, a disposable stepping stone as humans colonize space; humans should abandon and transcend our birth planet through uploading our brains to machines, or riding rockets to colonize other planets. Fatalism increasingly inflicts our lives with ecological anxiety and climate grief, leaving many horrified and paralyzed by a vision of Earth-as-Corpse/Graveyard.

These narratives are unsatisfactory. My goal in writing this book is to explore a different set of narratives for humanity's relationship with the nonhuman, both natural and digital: kinship, regeneration, collaboration. Instead of escapism, a new way to root ourselves in place, to feel fully at home on planet Earth. Instead of fatalism, I believe it is important to adopt a strategy of cautious experimentalism: modest attempts to restore habitability and reduce precarity of life for humans and nonhumans alike, in a spirit of solidarity. Kevin Kelley coined a term for this type of thinking: protopia—a thought experiment that emphasizes the positive potential of future pathways. As Gramsci once put it, pessimism is inevitable in moments of crisis, yet these moments also demand optimism of the will and creativity of the imagination.

If these technologies are captured by Big Tech and deployed in the service of platform capitalism, the Digital Earth agenda may contribute to a grimmer future of worsening socioeconomic and socioecological inequalities and harms. But as I explore in subsequent chapters, Digital Earth technologies could also contribute to urgently needed sustainability goals: protecting biodiversity, mitigating climate change, reducing the tsunami of waste generated by resource extraction, and enabling more equitable participation of local communities in environmental monitoring and conservation. As the following chapter explores, these issues are being urgently debated in some of the world's most threatened biodiversity hotspots, where digital technology is having a measurable impact on protecting endangered species while sparking a wide-ranging controversy.

PARABLE OF THE DIGITAL POACHER

In a forest in Cambodia's Cardamom Mountains, a tiger walks by a camera trap. An algorithm called Wildbook confirms the sighting of the tiger and identifies the poachers following its tracks.

A satellite locks onto the poachers and summons rangers to its defense. Following a route optimized by artificial intelligence, the rangers close in on the poachers, who turn and flee. Their faces are caught on camera, their voice prints recorded as they pass the geofence. Their gunshots are recorded by a hidden microphone, adapted from a discarded cell phone and perched overhead in a tree. A digital forensic algorithm confirms the weapons. The rangers have more than enough data to track them down, even as they try to disappear back into the city.

For the first time in centuries, the tiger population is growing. But there are more buyers than ever, and the price keeps going up. When the poachers return, they hack into the digital systems on which the rangers now rely. The geofence drops, the satellite signal is lost, the camera goes dark. The digital tracking chip the tiger carries in its shoulder, intended to make it easier to protect, now makes it easy prey.

3 FACEBOOK FOR WILDLIFE

Tanya Berger-Wolf didn't expect to become an environmentalist. After falling in love with math at five years old, she started a doctorate in computer science in her early twenties, attracting attention for her cutting-edge theoretical research. But just as she was about to graduate, she became obsessed with a topic that surprised her professors and even herself: zebras.

While still an undergraduate, Berger-Wolf began working as a research assistant at the ecology department, building computer simulations of wildlife populations.[1] She was intrigued by the fact that digital technologies and biodiversity were following exponential trends, but in opposite directions. While the digital sector was burgeoning, endangered species populations were crashing. And in contrast to the deluge of data she had experienced in computer science, Berger-Wolf was shocked at how little data existed about the world's most endangered species.

The animal that caught Berger-Wolf's attention, the Grevy's zebra, was known in antiquity as the imperial zebra. Used by Romans in their circuses, the largest of the wild equines, renowned for their elegant stripes and striking gait, the Grevy's zebras once roamed large expanses of East Africa in giant herds. Today, fewer than 1,000 zebras remain, crowded out by farmers' fields and cattle ranges, and still hunted for their skins and meat. By the time Berger-Wolf learned about their plight, scientists were predicting that the iconic species might die out within two decades.

Maybe, Berger-Wolf thought, she could apply her digital skills in a way that could help save the zebras. But the computer scientists she consulted were discouraging. As one told her: "You're smart enough to do theory; why do you want to do this applied crap?" Others advised her to transfer to computational biology. Berger-Wolf tried it out, only to realize that she didn't want to spend her time sequencing the human genome or creating more accurate models of the human brain. She wanted to do something that would combine ecology and computer science to help save the planet.

The field of research she imagined had no name, nor precedent. Undaunted, Berger-Wolf asked around: "Who is the best ecologist in the world?" She was given the name of Simon Levin, at Princeton University. Berger-Wolf wrote to Levin out of the blue with an unorthodox idea: they would invent a new field, called computational ecology—a research agenda as ambitious as computational biology, but applied to ecological issues. Levin accepted, and Berger-Wolf moved to Princeton. Even there, it was an uphill battle. Most ecologists saw computer scientists as mere coders. And most computer scientists saw ecologists as mere sources of data. Berger-Wolf couldn't convince her fellow computer scientists of her vision's potential. And the ecologists, although intrigued, were dubious: How, exactly, could computer science help save endangered species?

A FINGERPRINT READER FOR ZEBRAS

Zebras were her breakthrough. Shortly after arriving at Princeton, Berger-Wolf began chatting with Daniel Rubenstein, one of the world's leading behavioral ecologists. She explained the problem to him. Ecologists wanted the Kenyan government to adopt stronger protections, including a national park. But the Kenyan government demanded an accurate census before proceeding with regulations that would likely incite strong resistance from local communities. Here, the ecologists faced a catch-22. The usual census methods (catching the zebras and painting numbers on them or shooting them with anesthetic darts in order to embed electronic tracking devices) were expensive, traumatizing, and put the zebras at risk of infection. Given how endangered the zebras were, these conventional methods couldn't be used. But visual surveys were slow, expensive, and inaccurate; it could take up to half an hour to identify a zebra from a photo, and the people-shy zebras were notoriously hard to track. The herd's babies were dying at an alarming rate, yet without a census the government would not act. Unless something changed, the zebras were doomed.

How could computer science help? Berger-Wolf's epiphany emerged during a field computational ecology course that she and Rubenstein co-taught in Kenya. By then, she had met and married her husband—an ecologist with an interest in zebras. One night, Berger-Wolf overheard her husband joking with some local field biologists as they gathered to identify zebras from individual photos, a tedious task that would take many days. All they needed, they mused, was an automated method to help them identify and catalog individual zebras. It had been yet another frustrating day tracking the elusive animals, and Berger-Wolf overheard her husband say: "Why didn't I think of it before? All we need is a bar code reader for zebras!"

A light bulb went off for Berger-Wolf. Zebras did indeed have unique stripe patterns. Although a bar code reader probably wouldn't work, something like a fingerprint scanner potentially could. Theoretically, zebra stripes pose a similar conceptual problem to the identification of human fingerprints, as each individual zebra has its own unique markings. While biologists had long used these patterns (which they sometimes refer to as "body prints") to help identify individual zebras in the wild, the process had never been automated. Berger-Wolf began collaborating with her PhD student Mayank Lahiri to develop an initial iteration: StripeSpotter. Their goal was to create a free, open-source program to which anyone could upload a photo of a zebra's flank, for automatic identification. The algorithm is fairly straightforward: the zebra is assigned a "stripecode," which is then checked against the database. If the zebra is already in the database, it is matched to earlier photos; if not, it is assigned a new, unique identity. The AI-powered algorithm can identify individual zebras from a simple photo and has no problem handling images of different sizes, oblique angles, and over- or underexposed photographs. After testing it against thousands of zebra photos that she took from various angles, including hundreds of photos she took personally during several super-light airplane flights over northern Kenya, Berger-Wolf verified the accuracy of the algorithm. With enough photos, a complete census of the zebra population was now feasible.

Berger-Wolf then teamed up with Chuck Stewart, a computer scientist at Rensselaer Polytechnic; together with his student Jon Crall, they decided to develop a more comprehensive computer vision method for identifying patterned animals, which they called HotSpotter. The near-impossible task of identifying and tracking individual animals, which used to take weeks or months, now takes mere milliseconds. As reliable as facial recognition technology, computers could now identify zebras as individuals.

THE "MANY STARS" ALGORITHM

Berger-Wolf knew that her idea was highly scalable. In theory, the HotSpotter algorithm could be adapted and applied to many of the nearly 9 million known species on the planet. Moreover, manual photos could be augmented with photos by satellites or drones and crowdsourced from social media. In other words, it could become a low-cost face recognition reader for the world's wildlife. She was just missing one key component: a robust data management system.

A chance encounter led her to the solution. At around the same time Berger-Wolf had dreamed up HotSpotter, Princeton-trained physicist Zaven Arzoumanian developed an unexpected passion for endangered whale sharks. Whale sharks are mysterious creatures: although they are the largest living fish in the world (with adults longer than 40 feet weighing in over 20 tons), little is known about them—except that global populations have halved in the past few decades. Arzoumanian's interest was sparked when a friend, software programmer Jason Holmberg, had a numinous encounter with a whale shark while scuba diving. Holmberg began wondering about how the highly elusive fish might be tracked, and asked Arzoumanian—by then working at NASA's Goddard Space Center—for help.

How could they automate identification of whale sharks? The creatures have hundreds or even thousands of distinctive white spots on their backs, as unique as human fingerprints, but the subtle variation is hard for human eyes to parse and differentiate. After searching through the literature, Arzoumanian homed in on an innovative stellar pattern-matching algorithm that had been developed decades earlier for use by astrophysicists. When Princeton scientist Edward Groth first created the algorithm, he was trying to automate the analysis of the billions of stars revealed by the newly launched Hubble Space Telescope. The observable universe as a whole contains somewhere between 30 and 70 billion trillion stars; the sheer volume of information overwhelmed astronomers. Groth's algorithms solved this challenge; Arzoumanian realized that the algorithm, which analyzes patterns of pinpricks of light in the sky, could be used to identify individual whale sharks, whose dappled skins have unique patterns that are as intricate and unique as stellar constellations.[2] Holmberg and Arzoumanian tweaked the algorithm, and then reached out to whale shark biologist Brad Norman, who immediately saw the enormous potential in citizen science outreach: Norman's large network of amateur whale shark spotters could upload photographs, and the whale spotter algorithm could identify individual sharks in seconds.

For decades, Norman had been photographing and identifying each whale shark by eye, a tedious process that could take hours or even days; he was eager to try an automated method. He soon launched a global campaign. A year later, with the contributions of more than 5,000 citizen scientists who reported tens of thousands of whale shark encounters in dozens of countries, more than 6,000 individual whale sharks were identified, with a successful match rate of over 90 percent. The number of known whale shark gathering sites doubled within a few years—a finding that would not have been possible using manual observation methods.[3] Says Norman, "Whale sharks stopped being random animals . . . and became individuals with stories and histories and futures that are yet to be written. And that's what makes it so seductive as a citizen science project."[4] With the help of over 8,000 citizen scientists, and by scraping YouTube videos, researchers were able to identify over 12,000 individual whale sharks from over 75,000 reported sightings. Those data led the International Union for Conservation of Nature (IUCN) Red List of Threatened Species to reclassify the whale shark from vulnerable to endangered and to determine that the population trend was declining, rather than stable as previously thought.

FACEBOOK FOR WILDLIFE

Based on their initial success with whale sharks, Arzoumanian and Jason Holmberg founded a nonprofit organization, Wild Me. But they hit a stumbling block: just like StripeSpotter, their original algorithm didn't scale. However, Holmberg had created a robust data management layer, just what Berger-Wolf needed. The three innovators teamed up, and the constellation and StripeSpotter algorithms were replaced with the HotSpotter-based approach, linked with the Wild Me data management layer. They now had a system that could potentially catalog any living thing (anything, that is, with colors, stripes, spots, wrinkles, scars, or notch patterns that do not change as the animal ages).

Then, they embarked on an ambitious mission: cataloging the world's wildlife. The pent-up need for automated animal identification systems was staggering. Berger-Wolf's email was jammed with requests. Researchers sent in massive datasets of photos from all over the world. The huge amounts of data now available were both blessing and curse: the Wild Me team now had a massive cataloging problem, and it still wasn't obvious how conservationists could use the information. So the team, which had by this point attracted half a dozen engineers, created Wildbook, a version of Facebook for animals.[5]

The founders' vision was ambitious: blend wildlife research with citizen science and computer vision to accelerate zebra monitoring. They billed it as extinction-combating software. Their ultimate goal was to create a universal animal recognition algorithm that could identify unique individuals in every species on the planet, like a low-cost facial recognition reader for the world's wildlife. With such an algorithm, ecologists would be able to easily and automatically identify and track any individual living creature on the planet, from birth to death.

Several years later, Wildbook encompasses hundreds of species. (If you're a fan of whales or dolphins, check out Flukebook.) The social media side is the visible part of the website, where members of the public can visit and perhaps fall in love with some charismatic megafauna (or microfauna). At the back end, out of sight, is a set of algorithms that deploy computer vision, combined with citizen science, as a way to put a face, name, and story to each and every individual animal on the planet. The platform is now being used to track a Noah's ark–like list, including whales, giraffes, manta rays, humpback whales, Hector's dolphins, sea bass, flapper skates, turtles, sharks, jaguars, lynx, seals, polar bears, and sea dragons. With funding from the Moore Foundation, Berger-Wolf is expanding Wildbook to thousands of species. In the future, she hopes to develop systematic assessments of populations for every species on the IUCN Red List, a global inventory of endangered species.

Another funder of Wildbook was in part funded by Microsoft's new AI for Earth program. Why did Microsoft decide to get involved? In the words of Josh Henretig, who oversees the new $50 million fund Microsoft is investing in AI for Earth: "What's really worrying to us and to many scientists around the world is that we have only discovered/described about 1.5 million species of an estimated 10 million on our planet, and less than 5 percent of that 1.5 million species have ever been analyzed in any detail. There are species that are disappearing off our planet that we've never even known about."[6] Many researchers are now following in Wildbook's footsteps, developing machine learning algorithms to track specific species of wildlife (chimps, dolphins, badgers, birds, koalas, kangaroos), and even track the exotic pet trade. These efforts are not enough to stop extinction, but they are one important piece in the larger puzzle: accurate documentation of species decline.

In the meantime, the Grevy's zebras now have a fighting chance at survival. Soon after the launch of HotSpotter, Dan Rubenstein came up with a controversial suggestion: convince the Kenya Wildlife Service to incorporate HotSpotter into a biannual zebra population census. As Berger-Wolf remembers, "The computer scientists on our team nearly had a heart attack when Dan proposed a census." But Rubenstein insisted. They needed to convince the government to do a census, and they also needed to win over the public by engaging ordinary Kenyans in the process.

Although dubious, Berger-Wolf approached the Kenyan government, and to her surprise found some interest. After two years of negotiations and planning, the first two-day Great Grevy's Rally was held across Kenya in

2016. The rally was cast as a blend of citizen science and public relations, mobilizing hundreds of Kenyans, from the prime minister to children from the slums of Kibera, to snap photos of the elusive zebras in a two-day national campaign. The rally produced an unprecedented result: a verifiable, accurate census (rather than estimate) of the entire Grevy's population. Over subsequent years, its accuracy was confirmed; meanwhile, the political popularity of the zebra grew. As Berger-Wolf recounts, the youngest participant in the most recent rally was three years old, while the oldest was over ninety years old. The Kenyan government was able to enact reforms that were previously thought to be too unpopular to be feasible: passing a new Grevy's Zebra Endangered Species Management Plan (committing land, resources, and funding) and strategically limiting lion populations (a major zebra predator) with contraception. A government census in 2020 confirmed that—for the first time in years—the Grevy's zebra population in Kenya had stabilized rather than continuing to decline.

DIGITAL PARK RANGERS

Scientists are not the only innovators making use of digital technologies to track endangered species. In recent years, an alarming new trend has emerged: poachers using digital tech to track and kill animals with precision, further threatening endangered species around the globe. In response, conservationists have also begun exploring the use of digital technologies to detect poachers—catching them in the act, or even preventing environmental crimes before they occur.

Of course, poachers have long thwarted conservationists' best efforts to protect endangered wildlife. Although poaching is not the only cause (climate change, urbanization, pollution, and deforestation also play a role), it is still the biggest factor in the precipitous population decline of many highly endangered species.[7] Despite international laws like the Convention on International Trade in Endangered Species (CITES), the multibillion-dollar illegal wildlife trade is thriving and is the main driver of the catastrophic declines occurring in some species. Rhino horns and elephant tusks are the best-known examples, but poaching covers a much broader range of species:

lions, tigers, leopards, bears, turtles, otters, eagles, parrots, sharks, tortoises, anteaters, fish, and even plants. One in five vertebrate species is regularly traded internationally (for mammals, the proportion is one in four). Animals are traded both dead and alive, whole or dismembered, for everything from their skins (for display) to their organs (for traditional medicines).[8] The first—and often only—line of defense against poachers is park rangers. But park rangers are often faced with an impossible task: patrolling huge swaths of territory, with limited knowledge of where poachers (who are armed with increasingly sophisticated technology) might strike next. As conservation has become increasingly militarized, the job has also become increasingly dangerous, with hundreds of guards and poachers killed annually.[9]

If so, it will be partly thanks to the volunteer efforts of a Chinese computer scientist now living in Pittsburgh, Fei Fang. After graduating from Tsinghua University, she pursued her dream: studying computer science at Harvard. Fang's research deals with game theory in counter-terrorism and security strategy. Her first grad school project involved optimizing the patrol routes of Coast Guard boats protecting the ferries between Manhattan and Staten Island.[10]

When she started her PhD, Fang didn't plan on working on environmental issues. She was set on a stellar, if conventional, career in computer science. But then, out of the blue, she received an email from another graduate student, an ecologist. They had heard about her research on the news and had a wild idea. Could her work with the Coast Guard be adapted to help stop illegal logging? Fang was curious. She began reaching out to conservation agencies and attending environmental workshops, asking, "How can I help?" Few people had an answer. But in one conversation, someone mentioned tigers. After they explained the precipice on which the world's remaining tiger population currently stands, Fang was hooked.

A century ago, tens of thousands of tigers roamed the Earth. Today, only a few thousand remain.[11] Tigers are worth millions of dollars on the black market, driven in part by demand from traditional Chinese medicine. Although tiger bones and penises were delisted from the official Chinese pharmacopoeia in the 1980s, illegal trade in the products still flourishes. Fang, who had grown up in China, felt a personal connection to the issue.

Could she adapt her research to help defend the world's last remaining tigers from poachers?

Although it wasn't obvious to her at first, Fang soon realized that the needs of the Coast Guard had many parallels with those of park rangers. Both had a huge territory to cover, with limited resources; routes had to be selected to maximize the chance of finding terrorists (or poachers). And both guards and rangers had to worry about being attacked; randomizing their routes was a key strategy in minimizing their vulnerability. On this basis, Fang began developing a new algorithm. She christened it PAWS: Protection Assistant for Wildlife Security. Much like her earlier algorithm predicted where terrorists might strike Coast Guard ships, PAWS helps rangers and game wardens predict where poachers will appear next.[12] The algorithm uses a form of artificial intelligence called machine learning, combined with game theory, to predict routes of both poachers and protected animals. The AI algorithm first maps poachers' past behavior (e.g., snare locations) onto habitat and terrain to predict their routes. Once poachers' highest probability routes have been identified, PAWS uses game theory to randomize ranger patrols along those routes. As the guards patrol their routes, and report signs of poachers, the AI algorithm learns to better predict where poachers are likely to appear next. The result: park guards are more likely to catch poachers and less likely to be ambushed, because their routes are no longer predictable.

Less than a year after getting the original email, Fang flew to Malaysia to test PAWS in a protected conservation area. On its first trial, PAWS found a dozen elephant snares and caught its first poacher soon after. Next came Uganda. Then Cambodia and Siberia. Collaborating with her PhD supervisor Milind Tambe (director of Harvard's Center for Research on Computation and Society), Fang is now integrating her algorithm into the world's largest digital conservation monitoring network (the SMART Partnership); in 2021, PAWS was introduced in over 800 wildlife sanctuaries in over 60 countries. Microsoft engineers have also created an application programming interface (API) for PAWS, so it can be used by conservation organizations worldwide. According to Jonathan Palmer, chief technology officer of the SMART Partnership, "PAWS will deliver insights into protected areas around the world that weren't even possible to dream about five years ago." Already, Fang's algorithm has saved hundreds of animals—and their guards—from poachers.

Yet Fang, now a professor of computer science at Carnegie Mellon, is modest about her innovation and realistic about the challenges. Her algorithm can't tackle all of the threats to tigers, nor address the problem of ranger-poacher collusion. She anticipates needing to regularly update her algorithm, as increasingly sophisticated poachers collaborate with hackers to penetrate the digital networks increasingly used by park rangers. Her next step: incorporating drones and remote sensing into the rangers' software platforms, extending their ability to detect poachers before they strike.[13]

Similar systems are now being developed by other researchers.[14] RESOLVE, a conservation not-for-profit based in Washington, DC, is working with Microsoft and Inmarsat satellites to develop TrailGuard, which uses Intel chips in its cameras to carry out AI-powered image analysis locally, filtering pictures of human intruders, increasing successful detection rates. The beta version of the software, deployed in Tanzania in 2018, detected several dozen intruders and enabled rangers to make thirty arrests from twenty different poaching gangs and seize over 2,000 pounds of illegal bushmeat.[15] In another example, the Zoological Society of London, working with Google and Iridium satellites, has developed the Instant Detect system: using Google's AutoML (machine learning) system, it enables AI to recognize animals instantly from pictures transmitted from cameras via satellite. Acoustic sensors—which can detect sounds like chainsaws, engine noise, or gunshots—are also being used in poacher detection.[16] Rainforest Connection, a not-for-profit, uses recycled cell phones as acoustic sensors that it installs in national parks in Costa Rica, the Philippines, and Indonesia, "turning phones into forest guardians." When artificial intelligence algorithms used to analyze the recordings detect signs of poaching, rangers are alerted in near real time.[17] The hope is that these AI algorithms will provide protection to a significant proportion of the world's most endangered species.

POACHING GOES CYBER

Using digital technologies like AI to prevent poaching might seem promising, but the same technologies used to protect wild animals can also be used to locate and kill them with chilling precision. Animal tracking data can be misused by poachers, hunters, and fishermen; the same characteristics that render drones useful to park managers—inexpensive, discreet, accurate, and able to roam over long distances—also make them useful to poachers.[18]

In one incident, cybercriminals targeted the Panna Tiger Reserve in central India. Staff at Panna Tiger—which was recently named a UNESCO Biosphere Reserve—had successfully regrown their now healthy Bengal tiger population from just two adults to over fifty tigers. On the black market, the tigers were worth over $100 million. Each tiger was equipped with an Iridium GPS Satellite Collar, a device that cost over $5,000 each and is accurate within less than ten feet. The hackers attempted to access emails reporting the encrypted coordinates of an endangered Bengal tiger. Since the cyberattack, a team of wildlife officials now stays within a few hundred feet of the tiger at all times to deter poachers, aided by surveillance drones and wireless sensors that detect poacher intrusions.

Biologist Steven Cooke argues that cyber-poaching enabled by digital tracking devices is occurring with increasing frequency. Hunters in Wyoming have attempted to hack data on wolf radio collars in Yellowstone National Park.[19] Tags attached to great white sharks in western Australia were used to locate and kill the animals.[20] Computer scientists have tested one of the primary satellites used in animal tracking—the Argos satellite system—and found troubling security gaps, concluding that animals wearing trackers are now easy prey for cyberpoachers.[21]

Poachers are not the only ones taking advantage of digital technologies. In the past decade, poaching has gone digital, and, as with digital disruption elsewhere, poaching has begun scaling massively as a result. A global network of e-commerce sites has led to a boom in the online illegal wildlife trade, prompting the UN General Assembly to pass additional resolutions on illicit wildlife trafficking in 2015. China is the largest destination for illegal wildlife products and trafficked endangered species. The United States ranks second, and is also an important transit country for illegal trafficking from Latin America, with much of the trade destined for the exotic pet industry.[22] In addition to long-standing concerns over animal welfare and threats to biodiversity (from both poaching and the introduction of nonnative species), researchers have long been concerned about global disease epidemics caused by pathogens arising from human interactions with wildlife trade (of which COVID is a recent example).[23]

Over the past decade, Interpol and other international environmental organizations have documented an exponential increase in online trading in illegal wildlife over the past decade. In one survey, the International Fund for Animal Welfare identified more than 5,000 ads on more than 100 social media platforms and online marketplaces selling approximately 12,000 endangered and threatened species. The report is likely an underestimate, as it did not include password-protected websites, private Facebook groups, the dark web, or languages other than English.[24] In 2020, Chinese researchers published the results of a twelve-week survey of exotic pets advertised on four Chinese online platforms; they found 111 exotic species, 24 percent of which were mentioned on the IUCN Red List of threatened species, and 41 percent of which are banned under the Convention on International Trade in Endangered Species.[25] In a global survey published the same year, researchers concluded that the global wildlife trade also extends to a broad range of other species (including reptiles and plants). In the words of one group of researchers, the illegal wildlife trade now "permeates the Tree of Life."[26]

Concerns about illegal trafficking were already being raised two decades ago, as sales of ivory began appearing on eBay. The platform soon banned ivory sales.[27] But tech-savvy traffickers began using code words and emojis to describe ivory and other illegal wildlife products. Researchers began documenting how criminals were circumventing the law.[28] In response, researchers have begun developing machine learning algorithms to track the illegal wildlife and exotic pet trade online, in order to enable better detection and prevention.[29] Others have focused on predicting new trends, such as the rising popularity of pet otters on YouTube, or the emergence of "exotic animal cafés" in Asia, enabling early warning signals to be given to enforcement agencies and regulators.[30] Scholars in the digital environmental humanities have begun systematically investigating consumers' environmental attitudes and awareness online, particularly as expressed on social media; these "conservation culturomics" methods reveal new insights into consumers' motivations and consumption patterns.[31]

In the meantime, a new antipoaching algorithm was launched by Google and the World Wildlife Fund in December 2019. Wildlife Insights uploads images from camera traps automatically, and then uses an AI algorithm to identify the animal in the photo; the algorithm, which could recognize about 450 unique species, analyzed over 4.5 million images within the first month. The long-term goal is to aggregate footage from camera traps around the world to provide real-time statistics and trends on animal populations globally: a Waze for wildlife.[32] (To make sure poachers don't use the data for nefarious purposes, only vetted users will have access.) Data from Wildlife Insights is now being used to support the Coalition to End Wildlife Trafficking Online. To proponents of computational approaches, these methods seem promising. But skeptics argue that this is merely a digital version of whack-a-mole, as illegal wildlife traffickers are continually inventing new methods to evade detection. Without a comprehensive approach to safeguards and security, digital technologies may not be net positive for global biodiversity conservation—and could even accelerate biodiversity loss. Even if online abuses are regulated, the underlying factors driving biodiversity loss—such as urbanization and agriculture—will not necessarily be addressed. Indeed, recent assessments of biodiversity show worsening trends; the Convention on Biological Diversity recently announced that none of the targets set in 2010 have been reached.[33] It is both unrealistic and naïve to think that digital technologies alone will turn the tide.

COMPUTATIONAL SUSTAINABILITY AND CITIZEN SCIENCE

Berger-Wolf's and Fang's work exemplifies a new scientific discipline that has emerged in the past two decades: computational sustainability (also referred to as computational ecology).[34] Credit for coining the term goes to Carla Gomes, a professor of computer science at Cornell University. In

2008, Gomes proposed a novel idea to the National Science Foundation: a computer science agenda that would develop applications to support the United Nations Sustainability Development Goals. Gomes attracted a group of collaborators, including fellow computer scientist Tom Dietterich, one of the founders of machine learning. The group began mapping out environmental problems that could be addressed by computer science. Which were the most pressing conservation issues that could also be enabled by automation and enhanced data-intensiveness? Their attention soon focused on how novel applications of artificial intelligence could be used in biodiversity conservation.[35] A specific type of AI called machine learning is highly useful in classifying patterns in visual, spatial, and acoustic data, efficiently automating very large datasets. Moreover, machine learning uses data that can be gathered using minimally invasive monitoring.[36] WildTrack, for example, is a computer vision application that can identify endangered species from digital photos of their footprints, eliminating the need for tracking devices or camera traps. Automated systems can also enable environmental monitoring to continue even if researchers are on lockdown and national parks are closed (as was the case during the COVID pandemic).[37] And machine learning can also collect ecological data from nontraditional sources, such as social media.[38]

Much of this work is undertaken by scientists, but many Digital Earth initiatives also solicit the participation and collaboration of nonexperts as amateur "citizen scientists" through initiatives like volunteered geographic information (VGI) and the Participatory Geoweb. Citizen scientists may be involved in both crowdsourcing and cleaning data, and some initiatives are now very large; Cornell University's eBird platform—a citizen science bird observation community—has archived more than tens of millions of bird checklists and over a billion species observations.

Advocates argue that digitally enabled citizen science democratizes access to environmental data, yet critics raise privacy concerns and point to the lack of concrete conservation outcomes.[39] And the mobilization of digital tech in conservation presents other risks. Conservation agencies have a long history of conflict with local communities, particularly in low-income countries. This is in part because of draconian measures (such as evictions, arrests, restrictions of movement, and forced displacement) that are often used to establish and maintain parks and protected areas.[40] Digital technologies may exacerbate these conflicts. For example, some of the technologies described in this chapter (including drones) are used for surveillance of protected areas to exclude and track members of local communities and environmental activists. Digital surveillance may advance technocratic management, but it may also be deployed against grassroots resistance to resource and industrial projects, particularly in authoritarian states.[41]

A related concern pertains to the militarization of conservation.[42] Many digital technologies originated from research designed to support military and surveillance activities. There is ongoing debate within wildlife conservation about the increased use of techniques derived from military and security services—such as the development of informant networks or the use of counterinsurgency tactics against local populations.[43] State-sponsored surveillance under the guise of collecting environmental data might be mobilized to support state security objectives; the militarization of Digital Earth objectives might exclude citizens from accessing environmental data.[44]

Researchers have also raised concerns that some contemporary discourses about conservation conflate security and environmental concerns; in some cases, conservation agencies become use violent force against people they identify as poachers, counterinsurgents, or terrorists.[45] Some conservationists worry that certain actors engaged in conservation projects are actually actively serving members of intelligence agencies. Intelligence agencies have used nongovernmental aid organizations and vaccination programs as cover for clandestine operations and intelligence collection. For example, after a public outcry following revelations that a hepatitis vaccination program was used as a cover for the hunt for Osama bin Laden by the US Central Intelligence Agency (CIA), the US government announced that it would no longer co-opt vaccination programs in this manner, but no such guarantee has been extended to wildlife conservation programs. In this context, caution about digital technologies is merited: they may extend surveillance and enable or legitimate violence against local populations, without addressing the root causes of, for example, poaching.

These examples illustrate how Digital Earth technologies are double-edged. The same characteristics that make drones and AI useful for park guards also make them useful to poachers, as well as authoritarian governments. Online tools designed to monitor endangered species can be harnessed by profiteers trying to maximize their returns from the illegal wildlife trade. It is unclear whether conservationists or environmental criminals will win the digital arms race.

There is thus a tension at the heart of the Digital Earth agenda. On the one hand, Digital Earth technologies offer powerful and radically new approaches to conservation. Digital Earth technologies enable real-time situational awareness, monitoring, and responses to biodiversity threats, and specifically threats to endangered species. This redresses a major flaw in twentieth-century environmental governance, namely, the post hoc nature of both awareness and management. When acid rain monitoring began in North America in the 1980s, data was sent by fax from individual factories to the Environmental Protection Agency, with lag times of days, weeks, or even months; today, analogous data is available to regulators near-instantaneously. By providing hyperabundant and real-time data, the Digital Earth tech stack offers potentially powerful ways to address some of the most intractable problems in biodiversity conservation. But these technologies are easily misdirected toward surveillance and militarization goals, with negative impacts for local communities. And while Digital Earth technologies might aid the efforts of individual conservationists, they have not yet addressed the structural drivers of biodiversity loss.

But could they? In the following chapter, we will explore how Digital Earth technologies might address such structural drivers, including perhaps the greatest threats of all: climate change and human encroachment on the biosphere.

PARABLE OF THE ICE WAYS

Outside of Qikiqtarjuaq (population 598), an Inuit hunter drives his snowmobile due north. Behind his snowmobile, he pulls a qamutik, a traditional sled, fitted with sensors that monitor real-time ice conditions passing under the wooden runners. As he travels, he stops and inserts other sensors, built by the community's teenagers, into the ice.

The data will be relayed to a satellite that is tracking the ice that, as the Arctic warms, is shifting faster than anywhere else on Earth. The data will be housed in the Inuktitut language on Siku, an Inuit-built app that combines real-time maps with traditional knowledge, informed by satellite and sensor data. The digital maps help the hunter stay safe in a world where the climate is changing faster than anywhere else on Earth: an Indigenous Waze for the newly unpredictable Arctic ice.

Inuit *Qaujimajatuqangit*—the ways of knowing the land's being—is evolving as it has always done, but the digital maps are not always accurate, making travel riskier. "Country food" is a vital necessity in communities this far north, but the animals are scarcer, warier, harder to reach. Digital apps may be useless in a future where the old ice ways melt away beneath your feet.

4 HACKING CLIMATE

When Amazon workers staged their first mass walkout in the company's history, they had an unlikely source of inspiration: sixteen-year-old Greta Thunberg. In the fall of 2019, Thunberg crossed the ocean in a sailboat, arriving in New York just before September 20, the day that millions of young people around the globe skipped school for the biggest climate change protest in history. Google searches for climate change soared to record highs (even briefly beating out queries about Donald Trump's impeachment). The climate walkout at Amazon was all the more remarkable given the company's relentless work culture, hostility to unions, resistance to public airing of internal debates, and relative reluctance to act on environmental issues.

Similar walkouts took place at Twitter, Google, Microsoft, and Facebook. Employees spoke out against the soaring greenhouse gas emissions from the tech sector and, in some cases, the close relationships among the companies, the fossil fuel industry, and climate change deniers. Labor organizers at the Tech Workers Coalition, an umbrella organization for tech sector unions, promised that this was only the beginning and that labor action at tech companies was a growing movement. Amid scandals over fake news and privacy, Silicon Valley executives now had climate change politics to worry about.

As criticism proliferated, tech companies jumped on the green bandwagon. Climate change was the theme for 2019's annual elite Google Camp, although Twitterverse scorn ignited as hundreds of private jets, Maseratis, and mega-yachts transported guests to the company's invitation-only luxury

Sicilian resort. Google's assertion of its carbon-neutral status did not assuage its critics. Meanwhile, just before his employees' walkout, beleaguered Amazon CEO Jeff Bezos announced a Climate Pledge. His employees argued it was too little, too late, and demanded that the company stop funding politicians who deny the existence of climate change.

As the year drew to a close, Microsoft president Brad Smith declared that environmental sustainability was *the* top tech sector issue for the coming decade.[1] In his annual end-of-year message, Smith argued that the tech sector needed to move more quickly than in recent years to improve efficiency and use more and better renewable energy. He went on to note that "this is just the tip of the iceberg": indirect emissions in the tech sector's value chain—such as those generated by the production of concrete to build new buildings, or by manufacturing new devices—had long been neglected. Other executives echoed his statement. But cynics lambasted Big Tech's environmental makeover as a self-interested attempt to salvage corporate reputations battered by scandals over monopoly power, privacy breaches, political manipulation, trolls, porn, and sex trafficking. Pundits mocked tech leaders' newfound interest in saving the planet as mere greenwashing, a thin veneer of environmental commitments masking a big-business-as-usual agenda. They also reminded the rest of us about the not-so-well-kept dirty secret of the global internet: it is a voracious electricity consumer.

Digital Earth generates a hyper-abundance of data with which to perceive the various facets of Gaia, the most scrutinized aspect of which is our global climate. The perception of our climate as a planetary-scale phenomenon, and the ability to monitor the climate with accuracy and precision, are products of the Digital Age. Beyond providing a deeper understanding of the Earth's climate system, Digital Earth technologies are also being used to create new strategies for climate action. In this chapter, we will explore both how growing Digital Earth technologies are furnishing new data sources and how applications of AI are being used to advance climate change mitigation (emissions reductions) and adaptation (responding to the risks and hazards of a warming world). The applications discussed in this chapter illustrate how Digital Earth governance is data-rich, predictive, and potentially useful in addressing climate change. We will also consider some of the ethical issues and risks posed by the AI for climate change, including its own high energy consumption and contribution to an increase in greenhouse gas emissions.

"OUR OWN DAMN SATELLITE": MITIGATING CLIMATE CHANGE

How could Digital Earth technologies be mobilized not merely to monitor greenhouse gas emissions but also to mitigate climate change in real time? Part of the answer lies in the recent explosion of satellite technology. Over 700 Earth observation satellites have been launched during the past decade. Many of these systems will be modular, made of dozens of satlets supplying everything a satellite needs—a source of power, data processing, mobility, and a way to communicate with Earth—at a fraction of the cost of traditional satellites. Even as they get smaller, satellites are becoming more complex, with infrared sensors that detect heat; hyperspectral sensors that identify specific vegetation, minerals, and other materials; and radar scanners that can recreate 3D landscape visualizations. The spatial resolution is breathtaking: even insects can be detected remotely.[2] And specific satellites can be developed for specific pollutants, such as carbon dioxide, methane, and carbon monoxide.

Frustrated by the US government's backpedaling on climate change action, then-governor of California Jerry Brown announced in 2018: "With science still under attack . . . we're going to launch our own damn satellite to figure out where the pollution is." The following year, the state of California announced a partnership with Bloomberg Philanthropies to launch a new Satellites for Climate Action initiative. The initiative focuses on three areas: analyzing emissions (e.g., from coal-fired plant operations), developing new satellite technologies that detect greenhouse gases, and developing new geospatial analytics that can enhance observation of forests, coral reefs, and species threatened by climate change. None of these satellite applications are inherently new from a technological standpoint, but the decision by state and private actors to launch their own satellites is potentially game-changing. The state of California has an ambitious plan to achieve carbon neutrality by 2045 while continuing economic growth; to reach this goal, cheaper and faster ways to reduce emissions will need to be found.

Bloomberg's involvement is no coincidence. Asset managers have begun using satellite data to predict physical risks to assets and infrastructure from climate change and extreme events. When BlackRock CEO Larry Fink declared in his annual letter that climate change would bring about a "fundamental reshaping of finance," CEOs took note.[3] But incorporating information on environmental, social, and governance (ESG) performance into investment decisions is challenging if the sole source of data is a company's annual sustainability report. By using satellite data to obtain real-time information on risks and emissions, investors have access to independent data to make investment decisions and push companies for change.

Meanwhile, environmental NGOs are also launching their own satellites. And they intend to do more than merely monitor: they will name and shame, and perhaps even sue, high-emissions perpetrators. The most high-profile example is the Environmental Defense Fund (EDF), which is partnering with Harvard University and the Smithsonian Institution to launch its own satellite, MethaneSAT. MethaneSAT measures and quantifies emissions from oil and gas fields, and also detects concentrated point sources of climate pollutants, while monitoring leaks from locations where emissions are known to occur. The specific target is methane, which is a significant contributor to global warming.[4] The EDF's stated goal is to lower methane emissions from targeted oil and gas systems by 45 percent by 2025. The satellite operates in target mode, focusing on 200 priority sites in the oil and gas value chain, many of which are known to have high levels of previously unacknowledged methane leaks.[5] Verification of these leaks can be achieved with helicopter or drone flyovers for spot checks.[6] And, unlike proprietary satellite data purchased by investors, all data will be publicly available.[7]

MethaneSAT signals the beginning of precision emissions regulation, and public shaming, for high-emitting companies. How is this different from previous approaches to emissions regulation? First, precision regulation is both spatially precise and scalable. Previous methods usually required a trade-off: you could either monitor things on a large scale or monitor them very precisely, but not both. Take, for example, the case of fugitive methane emissions from natural gas wells. This is a major, and often undocumented, source of carbon emissions. In the United States alone, fugitive emissions from the oil and gas industry are estimated to be over 10 million metric tons per year, representing approximately $2 billion in lost revenue.[8]

So-called super-emitter wells are responsible for most of the gas that leaks. But finding them is challenging—a twenty-first-century version of the proverbial needle in a haystack, given the millions of wells spread across the landscape. Before MethaneSAT, the only way to independently detect leaks was to visit the wells one by one, testing for gas leaks with an infrared camera. In 2016, the EDF enlisted scientists to fly over 8,000 gas wells across seven states.[9] The largest independent study of well methane emissions conducted to date, it took several months and cost millions of dollars. To do a similar study of abandoned wells in British Columbia, scientist John Werring trekked more than 8,000 kilometers over the course of several months, visiting over 1,600 well pads and documenting the systematic underreporting of fugitive methane emissions.[10] In both cases, the scientists assessed only a small fraction of wells. MethaneSAT could accomplish the same task, for all oil and gas wells, in a few hours.

Second, precision regulation provides information in near real time. Most environmental monitoring in the past suffered from a significant time lag. Acid rain reduction regulation in the 1990s is a good example. The Acid Rain Program, which featured sulfur dioxide emissions trading, was enacted in 1990 under the Clean Air Act Amendments (CAAA). The act instituted a nationwide cap on sulfur dioxide emissions, set allowances for emitters (such as coal-fired electricity plants) based on historical emissions, and then allowed companies to trade these allowances with other emitters if they increased their efficiency and no longer needed them. The Acid Rain Program was controversial (were companies just "buying the right to pollute"?) and market failures meant it was imperfect; the program was later replaced by a different regulatory regime. But there is some evidence that it was cost-effective, and it sparked a wave of innovation that reduced sulfur dioxide emissions and increased monitoring. A key gap was the long lag between pollution and punishment. Emissions were tracked hourly and sent in ASCII format on floppy disks to the Environmental Protection Agency (EPA) by courier (or in some cases reported by fax). The EPA collated the data, added up the net emissions at the end of the year, and generated an automatic penalty for anyone who exceeded their cap. Emissions were monitored continuously, but reporting was intermittent. Precision regulation minimizes or even eliminates the time lag between an environmental action (whether positive or negative) and consequence (whether reward or punishment).

Third, precision regulation is agile and ubiquitous. To continue the acid rain analogy: monitoring systems (which were expensive) were installed on the premises of only the largest emitters. In contrast, digital technologies can be everywhere, while focusing with intense precision on somewhere specific at the same time. Global coverage, yet with a high degree of spatial resolution, enables alerts to be sent in real time when, say, an earth-moving machine crosses the boundary of a national park or a fishing boat moves into a Marine Protected Area. And, in the case of methane emissions, a range of novel technologies enables companies to detect emissions before regulators or the EDF do (saving money while avoiding being named and shamed). Current technologies, such as laser spectrometers mounted on aircraft, might cost up to $250,000, but the new generation of monitoring devices (such as frequency comb lasers or tunable laser diodes) are orders of magnitude cheaper, and therefore attractive to both industry and conservation organizations.[11]

Precision environmental governance means that polluters will no longer have anywhere to hide. And if regulators don't hold polluters accountable, investors and markets will. Data on pollution and methane plumes can easily be turned into financial risk metrics. Now that major investors like BlackRock and market analysts like Morningstar have signaled their commitment to mainstreaming environmental sustainability, markets are pricing climate risk at the sectoral level. In some cases, climate risk can be priced for individual companies; for example, satellites could enable responsibility for greenhouse gas emissions to be attributed to individual companies and even individual sites. Other companies are now providing detection services for relatively small leaks.[12] When it comes to methane emissions, there will be nowhere left to hide.[13]

Precision regulation can also be used to reward carbon heroes. In 2019, scientists announced a new method for monitoring carbon emissions from tropical forests, which combined high-resolution airborne laser measurements of canopy height with high-resolution satellite images. They are now developing an operational carbon monitoring system that quantifies the cost of deforestation. If countries opt instead for reforestation, their carbon emissions reductions can also be monitored—and rewarded.[14]

FIRE AND ICE: ADAPTING TO CLIMATE CHANGE

In addition to driving climate change mitigation efforts, Digital Earth technologies also promise to assist with climate adaptation: managing human responses to climate change–induced hazards. A new generation of consumer-facing applications has been created that translate climate data to individual consumers, policymakers, governments, and regulators. In some cases, this information is used in policymaking; in others, it is intended to enable "social nudging" (which is a form of automated, digital manipulation akin to peer pressure) aimed at individual consumer behavior, guiding users to more environmentally sustainable actions.

Improving responses to environmental hazards is one of the most powerful examples of Digital Earth adaptation applications. For example, Planet—a San Francisco–based satellite company—has announced a new initiative to monitor wildfire risk from space using AI, creating real-time fire risk maps for emergency response.[15] Planet is betting that real-time satellite imagery will transform how forest fires are predicted, prevented, and managed. It recently launched the California Forest Observatory, an initiative in which Planet has partnered with Salo Sciences (a Stanford conservation tech spin-off) and Vibrant Planet (a sustainability tech strategy firm, run by Netflix's former director of marketing) to create an AI-powered observatory to exponentially enhance fire hazard prevention. The observatory has an ambitious plan to dynamically map forest structure and fuel loads down to the level of individual trees statewide, creating an unprecedented, continuously updated view of wildfire risk across California.

Many firefighting agencies, as well as forestry scientists and managers, already use satellite images, such as Landsat, which was a breakthrough technology when its first satellite was launched in 1972, but the images it captures are of a coarser resolution than Planet's, and with a greater lag time.

Planet's approach enables near-real-time analysis, which is what firefighters need. Fire management agencies have also used laser-powered monitoring devices like LiDAR (light detection and ranging), which uses light reflected from objects as a kind of visual sonar. When LiDAR is used at near-ground level by a drone or low-flying plane, it produces images with even better resolution than the best satellites; maps of the canopy can capture the details of the tiniest leaves and branches. These laser maps provide invaluable data on the height and density of trees, which is critical for mitigating the most intense and dangerous fires. However, these laser maps are expensive and time-consuming to produce; scanning all of California's forests on a daily basis would be impossible.

Planet combines satellite and LiDAR technologies to train an algorithm to generate high-resolution maps from lower-resolution satellite images. The AI algorithm can then be fed real-time images, even if lower-resolution, and create an instant, accurate map of fire risk and vulnerability that is continuously updated. The California Fire Observatory will use these maps to achieve two goals. First: to help firefighters figure out what to do when fires actually break out. If lightning strikes in a forest, the observatory can immediately analyze the topography and vegetation and identify the level of risk. A fire in a remote and rocky area might require only continued monitoring, while a blaze near a community in a dense forest might require sending in firefighters and air support. Optimizing resources in this way could help prevent fires from getting out of control, while enabling greater confidence in allowing periodic natural fire events in some areas. The second goal is to optimize fire prevention. The maps will be used to identify regions at highest risk of canopy fires, enabling teams to strategically target areas for small, controlled fires to burn away flammable undergrowth. Today, teams can also use the observatory's high-fidelity information for thinning and strategic logging. Of course, this technology alone won't solve California's fire-related problems; for example, it won't prevent fires caused by aging electricity infrastructure. But it may save many human lives.

California, home to Silicon Valley, might well be able to afford to build AI-powered algorithms to protect itself from extreme events. But what about other parts of the world? In particular, what about isolated or sparsely populated regions, or communities that don't have internet access? As fires are burning in California and Australia, the Arctic ice is also melting. For the Indigenous peoples of the North, like the Inuit, this creates an existential threat, as travel over ice is the only way to hunt throughout much of the winter. With widespread food insecurity, hunting is a crucial activity for these communities. But as the planet warms, the sea ice is thinner, forms later, and breaks up earlier than before; as a result, hunting practices have been significantly disrupted. And unexpected midwinter warming events, which turn sea ice into slush, create even more dangerous and unpredictable over-ice travel. Residents of northern regions feel that they can no longer predict ice conditions. With alarming regularity, travelers are falling through the ice, even in areas that were traditionally reliable. With elders and youth increasingly afraid to go out hunting, food security is threatened and traditional hunting knowledge in danger of being lost.

SmartICE integrates traditional Indigenous knowledge of sea ice with remote monitoring to create a novel digital climate change adaptation tool. How does it work? The system begins with Indigenous knowledge, which is based on millennia of observations of local ice conditions. Inuit elders inform decisions about locations for the deployment of a network of autonomous fixed sensors placed in the ice, which record ice thickness and other relevant data. Additional sensors are attached to sleds that the Inuit haul by snowmobile, which measure sea-ice thickness along travel routes. Data from the sensors, which are built locally, are networked with satellite imagery and cross-referenced with Indigenous knowledge to create sea-ice hazard maps, with user-defined categories, that distinguish safe and dangerous travel zones. The maps are updated weekly and disseminated through a cloud-based interface, a SmartICE app, and hard copies—still relevant, as many communities and most backcountry regions in the Arctic have intermittent or no internet access. The app thus augments Inuit knowledge to enable safe sea-ice travel, food security, and the preservation of Indigenous knowledge.[16]

Should apps like SmartICE be welcomed, or should digital innovation, even on climate change, be treated skeptically, given ongoing processes of colonialism and appropriation? Heather Davis and Zoe Todd observe an insidious asymmetry in the different ways Indigenous and non-Indigenous persons interpret the climate crisis. As they point out, many Indigenous peoples have already endured successive waves of harmful environmental transformations, stemming from colonialism, capitalism, and industrialization. Climate change is just one more shock wave. As Davis and Todd point out, the climate change threat is, in fact, not new; it has been building for centuries.[17] Climate scientists have recently demonstrated the link between the mass genocide that occurred in the Americas after the arrival of Columbus and climate cooling; the carbon dioxide reabsorption that occurred as forests took over lands previously cultivated by Indigenous peoples caused the Little Ice Age.[18] From an Indigenous perspective, the Anthropocene—which mainstream scientists argue began in the twentieth century—dates back to 1492.[19]

As journalism professor Candis Callison argues, climate change is a kind of colonial déjà vu; climate injustice is just a recent episode in a longer history.[20] Daniel Wildcat argues that failing to acknowledge that Indigenous peoples will be disproportionately impacted by climate change, but also have unique capacities and knowledges from which responses develop, further entrenches colonialism.[21] The field of Indigenous climate change studies seeks to document and support Indigenous peoples' capacities to address anthropogenic climate change. It includes an acknowledgment that Indigenous peoples have living knowledges and heritages of adaptation to environmental change, and documents the diverse responses around the world, from the "right to be cold" movement in the Arctic to self-determined Indigenous innovations. While digital innovations developed by Indigenous communities have proliferated, climate activism that overlooks issues of colonialism risks further entrenching inequality and injustice.[22]

The same is true in California. Preventive burning was widely used by Native Californians to manage and restore landscapes, but these practices were suppressed after colonization. Yet forests that have been managed in this way are less affected by large fires and regenerate more quickly. In northern Australia, the resurgence of Indigenous land management has included the incorporation of Indigenous knowledge into prescribed burning, significantly lowering fire risk; the catastrophic fires that raged in southern Australia in 2019 and 2020 were notably absent from large parts of northern Australia.

Some Indigenous communities are combining their traditional knowledge with digital tools: in Arnhem Land, Indigenous ranger groups use digital platforms like CyberTracker to geolocalize ecosystem restoration activities, including prescribed burns.[23] A revival of similar practices might be relevant in the California case, supported by Planet's maps. But the history and efficacy of Indigenous fire management practices is all too often ignored. Scientific amnesia, perhaps combined with a savior mentality, creates a pattern of overlooking knowledge that has been long held by Indigenous communities, and then claiming it as a "discovery" when the knowledge resurfaces. Indigenous scholars argue that environmental justice requires Indigenous data sovereignty, which includes community-based counter-mapping and acknowledgment of Indigenous ownership of data extracted from traditional Indigenous territories.[24] Otherwise, the Digital Earth agenda will turn out to be as extractive as any other resource sector.

HIDDEN EMISSIONS

At first glance, it may seem admirable to deploy Digital Earth digital technologies to enable climate change mitigation and adaptation. But digital technologies are often highly energy-intensive and, as a result, are themselves contributing to climate change at an accelerating rate. Debates about the energetic cost of the internet were first sparked by Harvard researcher Alex Wissner-Gross, who calculated that even a handful of Google searches used significant energy—equivalent to boiling a kettle to make a cup of tea.[25] Data centers have now grown to become some of the world's largest single users of electricity. The global information and communications technology (ICT) sector—which includes digital devices, data centers, and the networks that connect them—accounted for approximately 1.4 percent of global greenhouse gas emissions in 2020.[26] If the global tech sector were a country, its total power usage would rank third globally, just behind the United States and China.

How should these energy demands be met? On the one hand, digital infrastructure can be powered with renewable energy—in which case, the growth of the internet might actually accelerate our transition to a decarbonized energy system. If, on the other hand, digital infrastructure locks us into dramatic increases in the use of fossil fuels, climate emissions could accelerate. Could digital solutions support the integration of fluctuating renewable sources into energy systems, thereby offsetting increases in energy demand triggered by digitalization?

The tech sector itself has a role to play in transitioning itself and its supply chains to renewable energy, while figuring out how to increase the energy efficiency of digital devices. Some Big Tech companies, for example, have embraced renewables, either through purchasing renewable energy produced by third parties or converting their own facilities to renewable production. Apple, for instance, runs its entire global operation on renewable energy, and has promised to bring all of its suppliers onto renewables as well. Microsoft plans to be carbon-negative by 2030, and has committed to carbon neutrality over the company's entire lifetime; by 2050, it will remove from the environment all of the carbon it has emitted since Microsoft was founded in 1975.[27]

But beyond the most visible Big Tech companies, the tech sector's record on renewable energy is mixed. Video streaming companies, for example, are major emitters.[28] Cryptocurrency is similarly energy-intensive: Bitcoin-related emissions alone could push global warming above 2 degrees Celsius.[29] Researchers have found that the severity of Bitcoin's climate impact per dollar is similar to that of crude oil. Longer term, the shift to greener forms of cryptocurrencies may mitigate this impact. For example, Ethereum recently shifted from the conventional proof-of-work system to a less energy-intensive proof-of-stake system; an initial analysis found that this reduced energy consumption significantly.[30]

Environmental campaigners have taken note. Greenpeace's Clicking Clean reports have shone a spotlight on the tech sector's energy impacts.[31] By naming and shaming big brands, campaigning has triggered some changes, but the tech sector as a whole has done relatively little to counter its climate emissions. And little has been done to address e-waste, or the materials used in building digital technologies that also produce pollution. Some key components of digital devices—such as cobalt, tantalum, and lithium—are mined by hand in dangerous working conditions—at times using child labor—in desperately poor countries.[32] Ethical, efficient energy and materials use remains an existential challenge for the tech sector, notwithstanding the mainstreaming of the "Green ICT" agenda.[33]

THE DECOUPLING DEBATE

Despite these criticisms, some proponents argue that digital tech can make a significant contribution to cutting global emissions. The Global Environmental Sustainability Initiative, led by chief sustainability officers of some of the world's largest tech companies, argues that analyzing and optimizing material usage across sectors can make a significant contribution to reducing climate change emissions. Digital transformation—notably through automation and AI applications—of resource extraction, industrial manufacturing, energy networks, and agriculture could further reduce emissions in the future.[34] A collaborative assessment conducted by major environmental organizations, academic researchers, and ICT firms argues that the world's largest economies can halve emissions by 2030 and that digital innovation could directly enable a third of this reduction, and indirectly enable another third.[35] Because of the potential for digital innovations to drive efficiencies across sectors, most road maps for rapid decarbonization of the economy prioritize digital solutions, although protection of natural carbon sinks and transformations of certain sectors, like agriculture, from carbon source to carbon sink also plays a major role. Yet planned increases in the efficiency of networks and data centers will only partially offset the predicted dramatic rise in ICT energy use, and it is unclear whether energy use optimization will offset emissions more broadly.

This argument hinges on the question of whether digital technologies can help decouple energy use from economic growth.[36] According to this argument, digitalization is a key enabler of increased efficiency in both energy and material use; for example, the digitalization of energy-using products and services through Internet of Things sensors could optimize energy consumption. Similarly, a transformation in electricity networks through new sources of production such as wind and solar energy, as well through smart grid technology, will enable a reduction in emissions from electricity systems, currently the largest source of emissions worldwide. Digitalization is also a driver of the "sharing economy," usership-based consumption practices that are associated with dematerialization as society shifts from ownership to usership. For example, "mobility as a service," electrification of vehicles, autonomous vehicles, and digital traffic systems can reduce emissions from the transportation sector, while simultaneously reducing congestion and the need for new roads. Digitalization of supply chains, including the design of products for repurposing, sharing, reusing, and recycling, can reduce materials use. Energy usage in construction and buildings can be similarly optimized. In the future, artificial intelligence and Internet of Things technologies will help further optimize efficiency of products, services, and materials use, while also allowing greater traceability of materials and products, thereby supporting a circular economy in which waste is minimized, and hence energy use is further reduced. Some proponents also argue that social media platforms can help with climate advocacy and engagement; notwithstanding the dangers of misinformation and algorithmic propaganda, social media messaging and more subtle tools, such as social nudging, can also help shift consumer behavior.

The decoupling argument has generated intense debate. Some researchers have found that growth of ICT use is associated with increases in emissions, whereas others have found the reverse. Some evidence suggests that the relative rate of emissions growth is higher in developing countries.[37] Although digitalization may increase energy efficiency directly, and indirectly by driving economic change related to the rise of ICT services, it also contributes to economic growth as well as increases in materials usage for ICT-related products and supply chains. If the emissions from the latter trends are larger than the reductions due to the former, ICT will be associated with an overall increase in emissions.[38] Researchers have also questioned the dematerialization hypothesis, as recent evidence indicates that the rise of ICT in Europe over the past two decades did not lead to dematerialization.[39]

Others argue that the decoupling argument misses the point altogether: even the most rapid decoupling rates observed to date, if scaled to the entire planet, would be insufficient to achieve greenhouse gas emissions targets. Digital technologies might be a necessary tool for sustainable consumption and minimizing greenhouse gas emissions, but they are not the sole solution; strict enforcement of reductions targets will be required. While this debate is unlikely to be resolved in the near future, its latest iteration has taken on a new focus: artificial intelligence.

ARTIFICIAL INTELLIGENCE AND THE FUTURE OF ENERGY

Proponents of digital technologies, and their potential role in contributing to decarbonizing the economy, often refer to artificial intelligence as game-changing.[40] AI could, they argue, improve the performance of Earth system modeling. For example, AI could accelerate the identification and analysis of teleconnections—recurring, large-scale patterns of climate and weather anomalies that span vast geographical areas—which are often difficult to discern due to complex feedback loops. AI could also be used to provide more precise, real-time warnings of extreme weather events. Researchers have proposed various methods for advancing AI-informed climate modeling, including self-learning neural Earth system models, while also developing methods for using AI to strengthen climate change governance and enable the alignment of climate policies within and between countries.[41]

When described this way, these innovations sound promising. But from an environmental perspective, there is a drawback: AI algorithms are often highly energy-intensive.[42] One specific kind of artificial intelligence—machine learning—is increasingly woven into all aspects of our daily lives: from chatbots and digital assistants to movie and music recommendation engines. Training a single machine learning algorithm consumes significant amounts of energy, as each algorithm typically needs to process millions of images or data points as it is being trained. Take, for example, language processing AIs like the kind that powers ChatGPT. These AIs are trained via deep learning (a subfield of machine learning), which involves processing billions of articles and books. Computer scientist Emma Strubell assessed the energy consumption of four neural networks used in natural language processing; training one algorithm created emissions of nearly 600,000 pounds of carbon dioxide equivalent, about five times the emissions over the lifetime of an average American car.[43] The latest generation of generative

AI algorithms such as ChatGPT are highly energy-intensive to train, and their incorporation into search engines, such as Microsoft's Bing, Google's Bard, and Baidu's Ernie, is likely to dramatically accelerate energy emissions. According to the International Energy Agency, data centers—where generative AI algorithms are trained and used—already account for over 1 percent of the world's greenhouse gas energy emissions.[44] And although data center electricity usage could be optimized, the rapid scaling of electricity use due to generative AI is likely, at least in the short term, to outstrip efficiency savings.

As Larry Lohmann observes, it is as if the heat engine—the linchpin of nineteenth-century mechanization—has been harnessed to the Turing machine, the archetype of mechanical computation.[45] Fossil fuel–powered mechanization once enabled factories to extract more value from labor in less time. Today, electricity-fueled automation, much of which is still fossil fuel–powered, enables technology companies to extract more value from data in less time.

This may be a harbinger of things to come. If AI is to achieve most of the goals outlined in the previous section, then abundant data is a prerequisite. Should this trend continue, AI could potentially be a significant contributor to climate change. As a result, AI algorithms can be costly to train both financially—due to the cost of hardware and electricity or cloud computing time—and environmentally, due to the carbon footprint of the specialized processing hardware used for AI applications. Moreover, the use of AI in other sectors can inadvertently increase greenhouse gas emissions; this "rebound effect" occurs when machine learning increases the efficiency of a service. Machine learning, for example, can enable autonomous vehicles and enhance their fuel efficiency, but is projected to increase overall energy use—and hence emissions—if autonomous vehicles are not shared. While use of AI can lead to greater understanding of climate phenomena on the part of the general public as well as scientists, greater precision in the estimation of carbon emissions, and even enhanced efficiencies in energy use, our increasing use of AI will likely accelerate overall emissions.[46]

In response, leading artificial intelligence researchers have set out an agenda for using AI to combat climate change, arguing that machine learning could significantly reduce emissions, and also enable society to adapt more smoothly to a changing climate.[47] A range of AI applications could be used to reduce energy use in electricity systems, transportation, buildings, industry, and agriculture; enhance climate prediction and modeling; model societal impacts; inform individual consumer behavior; and enhance policy, including carbon markets (Table 2.1). AI could enhance climate science by accelerating the speed, accuracy, and precision of gathering and processing data on key variables, such as carbon emissions and temperature; analyze the effects of extreme weather; and improve predictions of weather events such as droughts and floods, as well as future climate patterns. AI could also reduce energy consumption by boosting smart grids; optimizing transportation patterns of cars, shipping, and planes; enabling smart recycling; and enhancing performance in industry and precision agriculture. It could accelerate innovation for carbon capture and geo-engineering. And AI could also be used for social nudging to encourage consumers to make more climate-friendly choices, while raising awareness of the environmental and climate impacts of their behaviors.

While proponents remain hopeful, and point to the multiple ways in which AI could be used for environmental conservation, critics argue that the AI agenda has to date largely sidestepped the issue of sustainability by failing to systematically publish the energy consumption of algorithms. Any AI algorithm will have an environmental impact, but as the ubiquity and complexity of tasks provided to AI algorithms increases, energy consumption will also likely increase. AI researchers have proposed that the carbon footprint of AI model development and training be reported alongside other performance metrics, but this is not yet the norm. And although a new open-source tool, CarbonTracker, aims to predict the carbon footprint of AI algorithms, such tools are not yet widely used.[48]

Many of these advances are still hypothetical. To date, the most convincing application of AI has been the optimization of energy grids to increase their efficiency and hence reduce fossil fuel use.[49] Some estimates suggest that AI could enable a significant reduction in global carbon dioxide emissions if systematically applied to strategic sectors such as agriculture, energy, infrastructure, and transportation. Google's DeepMind AI lab has used AI to predict wind farm energy output, enabling energy deliveries to the grid to be optimized.[50] By applying DeepMind to its own energy consumption problem, Google cut the amount of electricity needed to cool its data centers by 40 percent.[51] Similarly, researchers are now exploring how machine learning could drive energy efficiency in 5G networks.[52]

The Montreal Declaration for a Responsible Development of Artificial Intelligence, published in 2018, with Turing Award winner Yoshua Bengio as one of its coauthors, signals an aspiration for environmental sustainability that AI has yet to achieve. In calling for AI hardware to maximize energy efficiency and mitigate greenhouse gas emissions over its own life cycle, reduce electric and electronic waste, and minimize impact on ecosystems and biodiversity, the declaration puts its finger on the contradictions at the heart of the digital sustainability agenda and highlights a major oversight in the current AI research agenda. As Mark Coeckelbergh, an AI ethicist, points out, these issues are overlooked by most AI researchers.[53]

The Digital Earth agenda is thus much more ambivalent than its supporters might admit. The high energy consumption of current AI technology, and digitalization more broadly, tends to increase energy demands both directly and indirectly. Unless digital transformation is directed to decarbonization, the Digital Age will continue to accelerate climate change. Addressing this issue requires broader thinking, beyond energy systems themselves. How could digital technologies drive energy and resource efficiency? And how can industrial policy and political economic arrangements be reoriented toward an overall reduction in fossil fuel use, rather than fueling growth in conventional, energy-intensive modes of resource extraction and consumption in the name of entertainment, convenience, and profit? The next chapter turns to these questions, situating digital technologies within a broader debate about the sustainability of economic growth.

PARABLE OF THE ROBOT MINER

Near the tiny Pacific island of Nauru, a remote-controlled robot crawls the seafloor, drilling for cobalt. The autonomous robot is monitored by a team of engineers scattered all over the world, working round the clock. A remote-controlled submarine brings the precious metal to a nearby ship. Once full, the ship will travel to Shanghai. The cobalt will be embedded into batteries for electric cars from Seattle to Seoul, Berlin to Beijing.

The islanders warn of the destruction of the ocean floor. Phosphate mining has stripped Nauru bare, leaving limestone pinnacles fifty feet high. Most marine life has died, and all food is imported. No streams or rivers remain, and the mining boats bring drinking water to the island as ballast. From the shore, protestors launch tiny boats. But they have no effect on the automated miners, and the protestors return at nightfall.

Offshore, the robot continues drilling in the darkness. As noise cuts like a knife through the water, the deep-sea life is slowly covered in fine dust.

Never sleeping, never ceasing, the robot's sensors catalog the sacrifice zone.

5 A DIGITAL GREEN NEW DEAL

Humans are a litter-making species. While much has been made of our tool-making capacities, it is our ability to create cumulative garbage heaps at a planetary scale that is perhaps our most defining legacy. Over the past century, humans—mostly those in the wealthiest countries—have produced approximately 30 trillion tons of stuff: buildings and bridges, shopping malls and ships, cars and computers, tools and toys.[1] This translates into 50 kilograms (over 100 pounds) for every square meter of the Earth's surface. The scientific term for this massive stockpile is the "technosphere." By contrast, the collective biomass produced by living things other than humans weighs in at just 3 kilograms (less than 10 pounds) per square meter of the Earth's surface. Human detritus now outweighs living things by a factor of 16 to 1.

As environmental scientist Karen Holmberg argues, humans have become the modern equivalent of volcanoes, spewing waste and pollution in an unending industrial eruption.[2] Today's climate crisis is partially the symptom of a collective human disease: planetary-scale littering and hoarding. Looking back through Earth's long history, only the massive asteroids of the Cretaceous, or the giant volcanoes of the Permian, have made a similar impact. The technosphere requires massive amounts of energy, largely from fossil fuels, and produces astounding amounts of waste and pollution.

Humans are also a terraforming species. Over three quarters of Earth's ice-free land surface is now intensively used by humans, mostly for agriculture. Approximately 20 percent is relatively untouched wildlands.[3] As the human population has expanded, doubling since 1970, both ecosystems and wild populations have retreated as humans have cut down forests, planted crops, and built cities and roads. The amount of water impounded by humans behind dams is now so large that it measurably affects Earth's orbit.

Colonization is often defined as a political process entailing appropriation, dispossession, and genocide of oppressed peoples. But colonization is also a biophysical, metabolic process. Cattle replace buffalo. Wheat replaces prairie grasses. Towns replace forests. Scientists measure the biophysical colonization of the planet in various ways. One metric is called "appropriation of planetary net primary production," which is a measure of the annual productivity of the global biosphere. Global human appropriation of Earth's net primary production (which scientists refer to by the acronym HANPP) doubled in the twentieth century.[4] In simple terms, this means that humans are consuming ever-greater amounts of the raw resources produced by plant life on Earth, fueled by the sun, which form the base of our planetary value chain. As humanity's impact has grown, other species have been crowded out; for example, wild vertebrate species' population sizes have declined by 60 percent since 1970.[5] And researchers recently calculated that all of humanity weighs six times as much as all wild mammals.[6] Today's biodiversity crisis is the result.

Our success at terraforming and litter-making has a drawback: it reduces the likelihood that human economies and societies will endure. Our model of industrial production drives economic growth, agricultural production, the building of vast amounts of infrastructure, and the generation of ever-increasing amounts of pollution and waste. These, in turn, are drivers of both climate change and biodiversity loss—symptoms of an underlying phenomenon: the Great Acceleration, which is the term scientists use to describe the dramatic increase observed over the past century in human impacts on the planet, including rapid and in many cases exponential increases in water and energy use, carbon emissions, domestication of land, and terrestrial biosphere degradation.[7]

To describe this unprecedented situation, scientists sometimes refer to our current era as the Anthropocene, in which humans have become a driving force shaping environmental change at a planetary scale. A core driver of the Anthropocene is digitalization, which has accelerated global economic consumption and production, ratcheting up resource extraction, pollution, and waste—particularly in the world's wealthiest countries, whose collective consumption of resources, and contributions to climate change and pollution, far outstrip those of the rest of the world. Addressing the challenges of the Anthropocene thus requires us to analyze the digital dimension of human impacts on the planet, while being mindful of ecological justice issues such as socioeconomic inequality and overconsumption.

How might Digital Earth technologies enable us to address these issues? In the previous chapter, I explored—with significant caveats—how digital technology might enable new strategies for climate change mitigation and adaptation. In this chapter, I will examine how digital technologies might be used to reorient our industrial ecology (the relationships between the economy, society, and the natural environment) toward more sustainable ends: minimizing resource extraction and consumption so that humanity does not continue to poison its own nest. I will focus on three sectors—industrial production, intensive agriculture, and urbanization—that have significant environmental impacts, and which have been the subject of digital innovation, including precision farming, smart cities, and circular economies. But first let me briefly recap the book's argument thus far.

In the past decade, scientists and conservationists have adapted digital tools to achieve conservation goals; thousands of such innovations now exist. Digital environmental monitoring and decision-making platforms are operational on every continent, in every major biome on Earth. Repurposed cell phones, hidden high in the tree canopy in tropical forests, are detecting illegal loggers. Antiterrorism software is being used to help predict and prevent poaching. Artificial intelligence algorithms use facial recognition to identify individual animals—from zebras to whale sharks—helping to track members of endangered species.

Digital Earth technologies have several implications for environmental governance. First, environmental data is becoming super-abundant rather than scarce. Second, environmental data is becoming ubiquitous: automated sensors, satellites, and drones collect data continuously, even in remote places that humans find difficult to access, sensing and managing the environment everywhere, all the time. This creates time-space compression (governance is temporally and spatially ubiquitous) and time-space agility (governance is spatially and temporally dynamic). Third, rather than responding to environmental crises after they occur, digital technologies enable near-real time responses and may even predict hazards and catastrophes before they happen. Environmental governance can thus be preventive rather than reactive, and environmental criminals will find it harder to hide. Crowdsourcing and citizen science can be used to involve the public in conservation efforts; citizen science sites such as Zooniverse herald a resurgence of public engagement in science akin to the Victorian era. Although the incursions of Big Tech into this space are cause for concern and raise ethical issues such as algorithmic bias and an increase in climate emissions, the engagement of thousands of not-for-profit conservation groups in this agenda creates a possibility that digital environmental governance might evolve to be more inclusive, enabling new patterns of inclusion, subsidiarity, and solidarity. But there is an important caveat. Digitalization might not necessarily reduce the underlying drivers of biodiversity loss and climate change: the massive increase in resource extraction and industrial production, and equally massive increases in pollution and waste, which have occurred at a global scale over the past century—driven by the consumerist appetites of the wealthiest countries in the world. Will digital technologies address these underlying issues or, rather, make them worse?

A DIGITAL DEGROWTH FUTURE?

When venture capitalist Marc Andreessen announced that software is "eating the world," his words were understood to be metaphorical. The phrase became a meme, signaling the digital disruption of economies and societies. But what if this metaphor were literally true? What if using a digital device meant that you were interacting with the world in a biophysical sense? This is, indeed, the case. If you are reading this book on a digital device, you are making a small but measurable contribution to climate change. Every time you search on Google, watch a movie on Netflix, or post on social media, your digital device consumes electricity. As discussed in the previous chapter, the rapid increase in electricity consumption of our digital ecosystems is in large part due to energy-intensive artificial intelligence algorithms. Andreessen recently updated his quip: "Software is eating the world, and now AI is eating software."

As discussed in the previous chapter, AI will intensify the already high level of energy use by our digital ecosystems. And although some tech companies have made commitments to using renewable energy, digitalization has tended to increase our electricity consumption—and hence become a major driver of climate change. Our digital devices feel pristine, yet their components travel from mines and factories, and often end up in e-waste dumps, generating pollution and human rights abuses in some of the world's most impoverished regions. While it is clear that digital technology is currently part of the problem—accelerating resource extraction, pollution, and the technosphere—it is less clear whether it could be part of the solution.

Some argue that a "Fourth Industrial Revolution" will enable digital technology to drive a decoupling of economic growth and resource use. Innovations in digital technology, combined with biotechnology and advanced manufacturing techniques such as 3D printing, may enable some degree of decoupling through increasing efficiency in supply chains and product life cycles. A 2018 report by the Global Commission on the Economy and Climate (an international group of business leaders, economists, and government officials) argues that economic growth can and will continue, driven by rapid technological innovation, increased productivity and efficiency, and sustainable infrastructure investment.[8] The report, titled "Unlocking the Inclusive Growth Story of the 21st Century," argues that cleaner, climate-smart growth could generate tens of trillions of dollars of economic benefits; carbon pricing and climate-related financial risk disclosure could accelerate investments in sustainable infrastructure and generate tens of millions of new jobs. Some environmental economists forecast that, by 2050, decoupling resource use from economic growth will have been achieved; some types of renewables are, for example, already cheaper than fossil fuels. But will this stave off the biodiversity and climate crises? The current debate is largely polarized between ecomodernists and techno-optimists, both of whom argue that decoupling is possible and sustainable economic "green growth" desirable, and degrowth advocates, who argue that decoupling, even if feasible, would not happen fast enough to avoid catastrophic ecological tipping points.[9]

Even within degrowth circles, debates have arisen over the role of digital technology. Proponents of digital degrowth argue that a range of benefits—decarbonization, resource efficiency, lower climate change emissions, and enhanced ecosystem protection—could be achieved more quickly and easily with digital innovations. Critics argue that this amounts to mere wishful thinking, as the risks of digitalization outweigh the benefits. Digitalization will (and already has) driven market concentration, as they scale exponentially due to low marginal costs, creating network effects that increase the likelihood of monopoly power. Digital platform capitalism is thus a new form of monopoly capitalism.[10] Although digitalization could theoretically support reductions in resource use, competitive forces may create perverse incentives that increase rather than reduce net human appropriation of Earth's resources. And speeding up adoption of digital technologies may accelerate risks not only to the environment and climate change mitigation but also to social cohesion.

Could digital technologies help decouple economic growth from environmental impacts? Should degrowth be digitally driven or low-tech? And what does digitalization imply for social equity? Regardless of where you stand on these questions, digital technologies are likely to radically transform how we farm, shop, and consume in coming decades. Examples of digitally driven decoupling can be found in all sectors of the economy. From an environmental sustainability perspective, the applications are incredibly diverse—from greening resource production (digital mining, precision farming) to greening supply chains (eco-business, sustainable consumer behavior, online commerce, circular economies, electronic waste) to greening infrastructure (buildings, utility networks) and fostering sustainable living practices (e.g., smart cities, greening industrial production). This creates enormous potential for exponential efficiencies in resource use, production, and consumption. In turn, this creates new possibilities for reducing the amount of land, water, energy, and resources that humans use—which may, in turn, support many sustainability goals. But, as the rest of this chapter explores, these technologies also create new risks.

Precision Farming

Enhancing the sustainability of land use is a key environmental challenge for this century. Stabilizing emissions from agriculture is critical to meeting climate change mitigation targets, yet increases in agricultural production are required to feed a growing human population. Achieving this increase while simultaneously meeting biodiversity and climate targets requires a massive transformation of agricultural production and food consumption patterns. In the past, increasing production was accomplished through unsustainable practices like expanding agricultural land into natural ecosystems (such as tropical forests) and intensive use of pesticides and fertilizers. But in the future, sustainable agricultural practices will require reductions in fossil fuel and overall energy use, as well as increased efficiency per unit of agricultural area. During the 2022 UN Biodiversity Conference, countries agreed to a landmark commitment, known as 30 × 30, which aims to protect 30 percent of Earth's land and water by 2030, making the need for precision farming even more urgent.

Precision farming promises to increase yields while reducing environmental damage by making more efficient use of resources, limiting agricultural impacts on fresh water and ecosystems, and thereby potentially enhancing biodiversity and protecting ecosystem services.[11] This has two dimensions: digital technologies (robots, sensors, drones, satellites) and cognitive systems (intelligent and increasingly automated) that, combined, enable increasingly precise, customized agricultural practices. This "sustainable intensification" entails the use of digital systems to enable the precise applications of seeds, water, fertilizers, and pesticides aligned with local conditions (weather, soil quality) and specific crop needs. Digital agricultural systems are also used to facilitate planting, tending, and harvesting via the networking and automated, intelligent control of autonomous agricultural machinery. Intensive factory farms are systematically using digital tools (such as radiofrequency identification, or RFID, tags for electronic animal identification, and dosing feed systems), enabling optimization of farming and traceability along the food chain.[12]

Big data is a pillar of precision farming: instrumentation and monitoring of fields, animals, and surrounding landscapes via sensors, remote satellite sensing, and computer vision, combined with remote sensing data sets, supports the optimization of agricultural processes—such as harvesting methods and timing, soil cultivation, and fertilizer and pesticide application—thereby reducing harvest losses and improving food quality. The gains are sometimes dramatic, with significant reductions in herbicide and pesticide use.[13] The integration of big data with agricultural systems enables farms, production systems, and value chains to be networked, creating efficiencies throughout the food supply chain.[14] Digital tech thus can, in theory, increase market transparency and reduce information asymmetries by providing more data to farmers and consumers as well as producers.[15] Increased information could also drive more sustainable diets through social nudging and expanded product labeling.[16]

Precision farming is a logical continuation of the industrialization of agriculture that took place over the latter half of the twentieth century. The so-called Green Revolution increased productivity gains in specific crops, thanks to a combination of scientific research and innovation on crop varietals, increased fertilizer inputs, and mechanization—in many cases supported by agricultural subsidies. These advances led to declines in relative food prices and to decreases in food insecurity in some regions, but not all regions or farmers benefited equally.[17] Scientists point out that the Green Revolution saved an estimated tens of millions of hectares from being brought into agricultural production, a clear environmental benefit.[18] But the methods used to intensify agricultural production have a well-documented environmental cost, as pesticides and farming practices led to degradation of freshwater ecosystems and biodiversity loss.

Like the Green Revolution, precision farming promises to increase efficiency and productivity. But will this necessarily enhance sustainability? The business model associated with precision farming is often aligned with large-scale monocultures and a concentration of capital.[19] The main incentive for farmers to adopt precision farming is to be more efficient; investments in digital technology end up saving money in labor costs, fuel, fertilizers, pesticides. But this requires capital-intensive investments, which are not equally available to all farmers; precision farming is thus more prevalent in the Global North.[20] Smaller farms may merge into cooperatives to form machinery pools, but smallholder and subsistence farmers are not necessarily well placed to benefit. Even large farmers are wary of becoming dependent on major agricultural corporations. Farmers are no longer owners but mere renters of complex digital agricultural equipment, which raises costs and limits technological choices, exacerbating the dependency of farmers on tech companies; for example, restrictive digital rights management (DRM) is used to undermine the possibility of independent repair, as debates over the "right to repair" movement have revealed.[21] Data sovereignty and data security are additional concerns, given the increasing amounts of valuable data collected from farms. And precision farming is associated with rising electricity consumption and increased electronic waste. Proponents of precision farming argue that it need not result in monopolization. In theory, digitally supported precision agriculture enables landscape design for small-scale sustainable practices—such as mixed cultivation, organic farming—that are more ecologically compatible. Autonomous, intelligent, lightweight machinery could, in theory, disaggregate production and empower farmers, rather than accelerate corporate monopolies. In practice, however, this is not the trend we observe.

Currently, digitalization is being used primarily to further intensify industrial agriculture by cutting costs and improving efficiency. Although digitalization may also support sustainable agriculture, this is not the priority, and may even lead to counterintuitive results. For example, reducing costs associated with animal husbandry may fuel accelerating demands for meat-based products, increasing the overall environmental impacts of the agricultural sector. There is also an animal welfare impact, as digitalization tends to lead to rising average herd sizes and higher outputs per animal, increasing crowding in mega-factory farms.[22] Sustainable, ethical outcomes are thus not inevitable results of the digitization of agriculture.

More generally, precision farming illustrates a paradox first identified in the nineteenth century by a British economist named William Jevons. The Jevons paradox states that efficiency improvements will tend to lead to an overall increase in consumption. Why? The invention of more efficient engines enabled cheaper production and train transportation, sparking the Industrial Revolution. But this innovation increased overall fossil fuel use, because it catalyzed more travel and more automation of industrial production. The more efficient our machines, the less they cost, and the more we use—resulting in a net increase in consumption. In the future, super-efficient technologies might increase our use of energy and raw materials, rather than reducing them.

Beyond the issue of sustainability, precision farming also raises concerns about equity. Rates of digital uptake vary widely around the world. Precision agriculture is widely used in industrialized countries, but in emerging economies precision agriculture is largely used for agricultural products destined for export. Most farms worldwide are smallholdings dedicated to subsistence farming; smallholders farm over half of the world's arable land and produce most of the food consumed locally. Many do not have broadband access or capital to access seeds, fertilizers, or digital technologies. This raises a significant equity issue, as digitalization may be accessible only to larger farmers; high-tech, capital-intensive industrialized precision agriculture could intensify inequities between farmers. Small-scale farmers could nevertheless benefit, particularly from text message–based advisory services (e.g., weather forecasts, pest and disease outbreak advisory services), and mobile phone–based digital apps (e.g., uploading photographs for identification and treatment options for pests and to determine crop nutrient requirements).[23] Virtual farmer co-ops and digital ledger–supported land registries could potentially enable farmers to secure land tenure, achieve better prices for agricultural products, and potentially avoid the large-scale land grabbing that is likely to accompany the digitalization of agriculture.[24] Nature-based solutions are often touted as sustainability solutions, but nature-based solutions may, in fact, facilitate new processes of capital extraction.

Moreover, precision farming creates long-term risks for farmers and food systems. For example, an increasing amount of agricultural data is held by private companies, with little transparency about how it is accessed or used. Digitally driven intensification of agriculture may lead to an increase in environmental impacts. Although promising, the potential of precision agriculture—to reduce operating costs and increase yields and quality while reducing environmental damage in a socially equitable manner—has not yet been realized and is unlikely without substantial regulatory safeguards.[25]

Digital Cities

This century is the first in which more humans live in cities than in the countryside. Rapid human population growth in the next few decades will be associated with even higher rates of urbanization. Cities are, proportionately, resource-intensive; urban areas consume disproportionately large amounts of resources and are responsible for more than two-thirds of global energy demand and greenhouse gas emissions.[26] Increasing urbanization will, unless sustainably designed, significantly accelerate climate change and resource use globally.

Proponents of digitizing cities (often termed "smart cities") argue that digital technologies are the solution to many urban sustainability challenges. By interconnecting infrastructure with technology that observes people's movements and behaviors, and using data analytics, municipal decision-making and urban services can be optimized. For example, optimizing urban transportation can include traffic monitoring with cameras and sensors, real-time optimization of traffic flows (with apps like Waze), and automation of vehicles. Urban dwellings may be similarly digitally controlled via smart sensors and energy-optimized control of electricity-consuming devices. To achieve these goals, smart cities require spatially extensive cyber-physical networks that intensively monitor the urban environment. From an environmental perspective, this is potentially positive, as digitized cities are able to systematically monitor and reduce pollution, resource use, greenhouse gas emissions, and waste.

Digital technologies can also transform urban mobility. Creating alternatives to motorized public transport can enhance urban quality of life, reducing traffic density, accident risks, pollution, and commuting times. "Smart" transportation systems can, for instance, enable shared mobility, which involves using rather than owning one's means of transportation (e.g., a car, bicycle, scooter) or using ride-sharing and ride-hailing services such as BlaBlaCar, Uber, Lyft, and Didi Chuxing. The service platform Mobility offers seamless intermodal mobility services; UbiGo and Moovel are two platforms that offer mobility subscriptions that include rental bikes, car sharing, car rental, taxis, and public transit. Underlying these systems is a range of technologies (e.g., smart traffic control and electronic tolls, parking guidance systems, vehicle platooning), environmentally friendly mobility technologies (e.g., electric vehicles), data analytics for real-time traffic management (traffic forecasts for dynamic traffic management, road tolls), coordination technologies (e.g., networking between individual vehicles and transport infrastructure), and assessment and pricing of negative environmental externalities (e.g., emissions, loss of time due to traffic jams, land consumption). Combined, these technologies aim to optimize urban transport, resulting in fewer vehicles, less space devoted to transit and parking, and lower pollution. To date, however, uptake has been low, confined to wealthier cities, and does not yet show decisive improvements in environmental impacts at a globally significant scale. Moreover, these technologies might create perverse rebound effects: average distances traveled may actually increase, for example, if autonomous vehicle use becomes widespread.

Given that global technology companies are significant drivers of urban digitalization, critics have raised significant concerns about smart cities—notably, the risks of state and private surveillance and market concentration in the hands of private providers.[27] Positional and locational data are of particular concern. Moreover, smart cities threaten to widen the digital divide; not all citizens can afford access to the smartphones and smart devices required to access services in a digitalized city. Cybersecurity is yet another concern: security gaps increase with the size of the network, and vulnerability to hackers increases as cities digitalize their core services. These risks have led some cities to create municipally owned digital facilities (digital infrastructures, fiberoptic cable network, free WLAN, collective cloud systems, and public storage for open data). Proponents of degrowth argue that smart cities will not naturally orient themselves toward sustainability unless digital transformation is combined with an open democracy agenda, whereby cities, through open policymaking processes, enable democratic control of data produced collectively. Digital rights–based policymaking is a growing agenda for municipal governments, although relatively few cities worldwide have adopted a digital rights framework systematically.[28] Not all urban digital initiatives necessarily serve the common good.

A concern over democratic control of data raises a related yet more fundamental issue: Who should guide the digital cities agenda: technology companies or governments? The cancelation of Google's Quayside smart city initiative in Toronto in 2020—after the company was plagued by criticism—is an example of the limits of the digital city agenda in private hands. Google launched the initiative in 2016 to much fanfare, promising to build a digital-first city "from the internet up." But concerns over privacy and social nudging of consumers soon surfaced. Even tech entrepreneurs spoke out against the initiative. Blackberry founder Jim Balsillie called the Quayside project "a colonizing experiment in surveillance capitalism attempting to bulldoze important urban, civic and political issues."[29] Senior government official Ann Cavoukian (the former privacy commissioner of Ontario) resigned from the project's advisory panel, noting in her resignation letter: "I imagined us creating a Smart City of Privacy, as opposed to a Smart City of Surveillance." Venture capitalist Roger McNamee called Google's smart city initiative "a dystopian vision that has no place in a democratic society," warning that companies like Google cannot be trusted to safely manage the data they harvest from residents. Similar warnings have been sounded about smart cities in China, which have been closely linked to surveillance systems created by the state, aligned with its Orwellian social credit system.[30] It seems unlikely that democracies will accept these intrusive technologies, whether owned by governments or private tech companies, yet most municipalities do not have the means to create them locally. The pace of the smart cities agenda will be slowed down, and perhaps even halted, by these unresolved questions of digital privacy, security, and data sovereignty.

Digital Sharing and Circular Economies

The economies of industrialized countries entail something surprisingly unique in human history: people buying items and discarding them, even though there may be others willing to borrow, share, or trade. This is the underbelly of the Jevons paradox mentioned earlier. When the costs of production drop and prices drop, people's interest in reusing goods also drops—and waste increases.

Proponents of "sharing economies" seek to reverse this trend. In sharing economies, trading of used goods constitutes the bulk of economic transactions between consumers, which minimizes the need for new products to be manufactured, purchased, and (ultimately) discarded. Thrift stores are a time-honored example of this, but contemporary sharing economies increasingly use digital platforms to connect participants. Craigslist, Facebook Marketplace, Freecycle, and other peer-to-peer platforms are general forums for exchange. Community-specific sharing economy platforms have also boomed in the past decade. Poshmark, for example, is a social commerce online marketplace where people in the United States can resell high-end fashion items—clothing, shoes, and accessories. Tool-sharing platforms like the Good Neighbor and Streetbank enable neighbors to set up tool-sharing collectives. Consumers can swap houses (NightSwapping, HomeExchange) and borrow designer clothes (Rent the Runway), camping gear (Outdoor Exchange, OutdoorDivvy), high-end camera equipment (BorrowLenses), textbooks (Chegg), and even boats (Sailo, Barqo). Usership rather than ownership is the common mantra of these sites, which seems to prove a key point made by degrowth advocates: people are happy to own less, as long as it's convenient. Reducing and reusing goods is the means to an end: saving money and fostering a sense of community.

Critics argue that sharing economies are fringe marketplaces that will not grow to replace the mainstream economy. And if they did, this would stifle innovation: without the economic incentives created by competition for consumer dollars, much-needed improvements in our products would stagnate. With respect to software, this claim does not necessarily hold true. GNU/Linux and Wikipedia are examples of how noncoercive collaboration and cooperation on globally shared resources can create products that are innovative and reach a large number of users. It remains to be seen whether the same could hold true for goods like clothing and tools.

Sharing economies are one example of a much larger shift actively promoted by a growing number of economists and environmentalists: circular economies. Our current linear economies rely on a take-make-waste industrial extractive model. In contrast, circular economies rest on three principles: regenerating natural systems, recirculating products and materials, and designing out waste and pollution. Advocates argue that such design-keep-regenerate economies, if combined with renewable energy sources, can enable us to decouple economic activity from the consumption of finite resources, with both social and environmental benefits. Circular economies are thus a paradigm shift: rather than merely reducing the negative impacts of linear economies, we need to change our entire economic model and many of the core assumptions on which it rests. Waste can be recycled as nutrients or raw materials, renewable energy is "solar income," and human and natural systems are inextricably interlinked. Mining the technosphere (like garbage dumps and tailings heaps) for valuable rare earth minerals is an example of a circular economy innovation.[31]

In circular economies, technologists strive to avoid both pollution and extraction of nonrenewable resources from Earth. Technological systems can thus conserve and reuse, but not consume. The only consumption is of renewable resources, between living things; for example, food and biologically based materials (like wool or wood) can be used, but then feed back into ecosystems through processes like excretion and composting—regenerating living systems like soil, thereby providing renewable resources for the economy. Biological systems can consume and regenerate and are the primary source of growth in a circular economy. In making these arguments, advocates draw on long-standing theories of cyclicality, systems-thinking, and metabolism in both living organisms and machines. Contemporary variants of the circular economy argument include ecologist Janine Benyus's concept of biomimicry, architect Walter Stahel's "functional service economy," the "cradle to cradle" design philosophy of architect William McDonough and chemist Michael Braungart, Gunter Pauli's "blue economy," and theories of industrial ecology and natural capitalism.

This collective set of developments is sometimes termed a Fourth Industrial Revolution: a convergent set of biological and digital innovations that will generate a complete makeover of industrial production (and industrial metabolism) through automation, cognitive computing, the Internet of Things, and cloud computing–enabled data exchange in manufacturing technologies. Ecomodernists argue that these cyber-physical systems will enable decoupling, supporting continued economic growth as total environmental impacts shrink. Degrowth advocates argue that this is unlikely: even if feasible in principle, decoupling will not happen in the time frame necessary to avoid ecological tipping points, so alternative policy trajectories based on technological deceleration, redistribution, and material-energetic degrowth will be required.

Where do digital technologies fit within circular economies? Some proponents argue that digital technology is vital to a transition to a circular economy by enabling production that is highly resource-efficient, low-emission, low-waste, and embedded in eco-industrial networks that are automated, decentralized, flexible, networked, and intelligent. Long-lived products with entirely recyclable components would reduce both resource use and waste. This would enhance resource efficiency not only in production but also in supply chains, by radically enhancing transparency through tracking and control of entire value chains. Some proponents also argue that virtualization enabled by digital technologies will also dramatically shift fossil fuel and resource use.

In part, this depends on how these value chains shift geographically in the future; some argue that value chains will shorten and localize, thanks to the emergence of open-source business models in design and manufacturing, linked to networked micro-factories—"fab labs" and makerspaces equipped with machines (like 3D printers and computer numerical control [CNC] machines) that can generate prototypes and devices from digital software files. This could enable a "design global, manufacture local" trend, reducing the need for global transportation of manufactured goods; easy-to-transmit commodities (knowledge, design ideas) are transmitted globally, while heavy, expensive-to-transport commodities (machinery, building materials) are local and, ideally, shared, reused, and recycled. In opposition to the large-scale digital agricultural model discussed above, farmware cooperatives like Farm Hack or L'Atelier Paysan (in France) support small-scale farming by sharing designs for open-source agricultural machines. The WikiHouse project shares designs of dwellings with minimal environmental impacts. RepRap shares open-source designs for 3D printers that self-replicate. Open Bionics produces open-source, low-cost designs for bionic and robotic devices.

Critics point out that the materials requirements of the digital economy have led to increasing volumes of hardware and infrastructure. Computing centers, servers, and transmission networks offer one obvious example of where this has been the case. As computing devices proliferate and become ubiquitous (e.g., wearables, or smart textiles woven into clothing) these demands will continue to increase. The rapid rate of innovation makes it difficult to predict whether and when this will shift to reusable products and/or renewable inputs, but short-term trends do not seem positive for the environment. Most forecasts suggest increased use of metals, rare earth minerals, plastics, and glass; amounts of e-waste are also expected to rise substantially in the next few decades, largely due to growing digitalization of industrial production and consumer goods.[32] Will Digital Earth monitoring add to the tsunami of e-waste already being generated? Whether the digital optimization of industrial value chains can offset these increases remains to be seen.[33]

Within this debate, issues of social justice often tend to be overlooked. The egregious human rights impacts of mining rare earth metals necessary for digital devices to function are, by now, well known. The energy transition from fossil fuels to renewables will create increased demands for minerals, which might create new economic opportunities but will also intensify risks of human and environmental rights, particularly in low-income countries.[34] An estimated 40 million people worldwide are de facto modern slaves by force or by fraud, working in gold mines in Ghana, graphite mines in China, granite quarries in India, and logging in the Amazon.[35] As Benjamin Sovacool puts it, the use of forced labor and sometimes child labor in artisanal mines, notably in Africa, amounts to "subterranean slavery" supporting the sustainability transition.[36] This forced labor, mostly in the world's poorest countries, provides a hidden subsidy to users of digital devices. Protests by local communities over the expropriation of land, inadequate compensation, and the health and ecological effects of mining are often met with violence, which in some regions—particularly in sub-Saharan Africa—is sometimes state-sponsored.[37] The Digital Earth agenda rests on an often hidden foundation of environmental degradation, human rights abuse, and socioeconomic injustice. Western science and industrialized moderns often choose to forget where raw materials come from, a collective amnesia about the dirty side of the cleantech revolution.

A DIGITAL GREEN NEW DEAL

To sum up my argument thus far: Our planet is currently undergoing two exponential changes that have profound implications for the future of humanity, and indeed for life on Earth. On the one hand, rapid digitalization is transforming our economies and societies, while creating new threats (e.g., automated misinformation, algorithmic bias). On the other hand, rapid environmental change creates exponentially accelerating threats to human life support systems (such as the atmosphere), threatening both environmental and human health.

These two trends are rarely considered together. Yet there are several reasons why the confluence of digitalization and environmental change should concern us. Humanity is entering the Digital Anthropocene. As discussed earlier, the Anthropocene is the name given to our current era, in which humans have become a geophysical force at a planetary scale, causing earthquakes, disrupting the carbon and nitrogen cycles, and altering the global atmosphere. The Digital Age refers to the accelerating digital disruption and transformation of our social, economic, and political lives. Although the interrelationship between the two trends is complex, it may be summed up as follows: digital transformation is affecting the trajectory of environmental change, and, conversely, environmental change will shape the future trajectory of digital innovation.

As discussed in this and previous chapters, Digital Earth technologies have significant potential pitfalls, in terms of both environmental impacts (energy use and e-waste) and socioenvironmental justice. If left unchecked, digital disruption tends to accelerate climate change, pollution, and waste. It also, as discussed in this chapter, tends to exacerbate human rights abuse because of the political economy of mining and supply chains that provide the necessary components for digital technologies. Without concerted effort to constrain these negative sides of digitalization, it will lead to an acceleration of resource extraction, consumption, and waste, raising significant issues of human and environmental justice.

Given this situation, it might seem naïve to argue for a Digital Green New Deal. Indeed, some critics argue that we need less digitalization, not more. But my opinion is that carefully regulated digital technologies can and should be used to advance environmentalism. This digital environmentalism agenda could, if thoughtfully managed, reduce resource extraction and waste, accelerate action on climate change mitigation and adaptation, and widen our toolkit for biodiversity conservation.

The Digital Green New Deal rests on two innovations: hyper-abundant data and real-time regulation. To briefly recap: digital innovation means that environmental data is becoming super-abundant (both spatially and temporally) rather than scarce. Automated sensors, satellites, and drones collect data continuously, even in remote places that humans find difficult to access. This creates the possibility of ubiquitous environmental governance: sensing and managing the environment everywhere, all the time. Moreover, rather than responding to environmental crises after they occur, digital technologies potentially enable us to respond in real time, optimizing efficiencies in resource and energy use, and sometimes predicting environmental hazards before they happen. If environmental governance can be responsive in real time, and be predictive, we can invent new ways of mitigating and preventing environmental harms that are preventive rather than reactive.

It is because of these two innovations—abundant data and real-time responses—that Digital Earth governance has significant potential to improve environmental monitoring, conservation, and decision-making. Whether this potential is realized depends on the broader political economy of digitalization. Digital technologies might create a virtuous cycle, whereby better information on environmental conditions and impacts informs choices about resource extraction, industrial production, and consumption; but they might also create a vicious cycle, further accelerating already unsustainable levels of resource consumption.

Digital Earth offers us two potential pathways: intensifying current centralized, large-scale, global industrial models of production, aligned with platform capitalism, or creating new decentralized, small-scale, local production models, aligned with community-controlled economies. The first path may deliver faster results in the short term but is material- and energy-intensive and poses significant risks to privacy, dignity, and security. The second path will be slower to deliver results, but digital conglomerates, under the aegis of platform capitalism, seem unlikely to create a more socially and ecologically just future. It is not the digital nature of the technologies that is critical, but the systems of ownership, enterprise, and governance in which they are enmeshed. If they ignore these issues, advocates of digital environmentalism are likely to unwittingly recreate the unjust patterns of the past, in which sustainable development becomes a strategy for capitalism to turn a threat into an opportunity, absorbing environmental externalities as new sources of profit.

The internet is the largest single thing that humans have built in our history. Millions of miles of fiberoptic cable now encircle the globe, expanding exponentially, the central nervous system of our economies, polities, and societies. The impacts of digitalization on our planet are simultaneously sociopolitical, economic, and ecological. As Gaia becomes overlain by a set of technical layers, our digital networks become more enmeshed in global ecologies, and global ecologies are more deeply influenced by digitalization.

This chapter concludes the first part of the book, the theme of which was regeneration: using digital technology to mitigate and adapt to climate change, restore biodiversity, and reduce waste and resource extraction. Thus far, in describing Gaia's Web, I have focused on digital technology as it is conventionally understood today. But digital technology, as I will explore in the chapters that follow, is evolving quickly. Biodigital technologies that use digital techniques to rewrite DNA, the code of life, enable scientists to manipulate organisms and entire ecosystems, create hybrid entities such as biological robots, and fuse humans, nonhumans, and machines. Biodigital convergence—the merger of digital and biological innovation—will have significant implications for environmental governance. In the second part of the book, I focus on the implications of biodigital innovation for environmental governance: multispecies environmental regulation (chapter 6), environmental rights for nonhumans (chapter 7), the cultivation of empathy through augmented and virtual reality (chapter 8), the use of biological robots or biobots as environmental sentinels and tools (chapter 9), and the emergence of biodigital forms of computation and dwelling through biological computing and AI-designed organisms (chapter 10).

It may seem difficult to imagine a biodigital future. Yet this is an important thought experiment. Whereas today's digital technologies are deployed by humans, tomorrow's biodigital technologies will be deployed by both humans and nonhumans. Today, computation is associated with inorganic machines and silicon-based innovation; but computation is widespread in nature, and scientists are creating machine-organism hybrids that exploit the innate computation mechanisms in living cells. My discussion of these innovations is inevitably speculative, but in each case I begin with actual examples that are being developed in the lab or deployed in the field.

The theme of the second part of the book is instantiation, a term that has a double meaning. For philosophers, to instantiate means to bring an idea into being. For coders, instantiation means the creation of a real instance of an abstract idea, such as a class of objects or a computer process. In using digital technology to address today's environmental challenges, we may choose to regenerate the Earth. In using digital technology to create new relationships with nonhumans, we may choose to instantiate a multispecies future.

II INSTANTIATING

Tell me what you build and I will tell you what cosmos you inhabit.
—Frédérique Aït-Touati[1]

PARABLE OF THE WHALE SINGER

Off the California coast, a pod of humpback whales is traveling south, following their ancient migration route to their birthing grounds.

A hydrophone near Santa Barbara picks up their songs. A satellite locks onto their location as they near Ventura. An algorithm broadcasts a message to the ships' captains sailing along the coast.

Slow down. The captains obey.

This is the first year in living memory that no whales die from ship strikes as they swim past Los Angeles. Even the largest ships move aside. The cargo waits, the passengers wait; the movements of the fleet are guided by the singing whales. But the noise of the idling ships often drowns out the whale song. In an ocean full of machines, few whales can be heard above the unceasing din.

6 A PARLIAMENT OF EARTHLINGS

On a chilly morning in January 1952, Alan Baldridge witnessed a murder. Sailing off the coast of Southern California in pursuit of a pod of migrating whales, he heard screams in the distance. The pod abruptly vanished. Scanning the horizon, he spotted a large gray whale swimming vertically and raising its head above the surface, a behavior known as "spy-hopping." Baldridge, a marine biologist at Stanford University, decided to investigate; drawing closer, he saw seven orcas singing hunting cries while devouring the lips, tongue, and throat of a small whale calf. Nearby, the baby's mother watched helplessly.

Baldridge's story inspired a controversial research agenda.[1] Soon after his encounter, the US Navy began using orca sounds in an attempt to control cetaceans. One of the navy's first experiments involved sailing a small catamaran off the coast of San Diego, playing recorded orca screams to gray whales swimming south on their annual migration. The results were, as the researchers noted, "spectacular": the whales whirled around and fled north, or hid deep in nearby kelp beds, slowly popping their heads above the surface to search for predators. When they finally resumed swimming south, the whales were in stealth mode: sneaking past, with little of their bodies showing above the surface, their breathing scarcely audible.

The navy's next experiment was in Alaska, where a local fisheries official named John Vania was at war with beluga whales along the Kvichak River, home to the largest red salmon run in the world. While bears and eagles feasted on the shore, belugas would surf the mighty tide up the muddy brown estuary, feeding on the endless conveyor belt of salmon swimming toward the sea. After the fishermen complained that belugas were eating too many fish, Vania tried chasing the whales with motorboats, blaring rock music, even throwing small charges of explosives—all in vain. But when he pumped the navy's orca recordings through jerry-rigged underwater speakers, every single beluga immediately turned and fled, battling against tides that were strong enough to fling large boulders into the forest. Vania soon discovered that although they responded to hunting screams, the belugas appeared most frightened of orca "clicks," as if a warning was encoded in the staccato sounds.

At the time, industrial whaling and dolphin hunting were still permitted. Whalers killed tens of thousands of bowheads, sperm whales, and right whales annually, and rendered their oil into lubricant, perfume, and lipstick. In pursuit of tuna, fishermen killed an estimated 6 million dolphins in the Eastern Pacific in a few short decades following World War II. But a ragtag organization named Greenpeace was beginning to send protestors on small Zodiac boats into Alaskan waters to protect whales from bullets and harpoons. Their efforts, caught on camera, inspired public outrage. A global movement to save the whales led to a commercial moratorium on industrial whaling, as well as to the adoption of "dolphin safe" fishing legislation.

When no longer permitted to kill cetaceans, fishers began using acoustic deterrents; the devices were required by national governments to be installed on fishing boats, fish farms, and even fishing nets. A truce of sorts was declared between cetaceans and humanity. Yet the apparent benefits were short-lived. Acoustic deterrence creates damaging side effects, including hearing impairment. In the underwater world, where sound travels faster than it does through air, cetaceans use echolocation (also called biosonar) to "see" the world through sound. When the din of motors and seismic blasts is added to acoustic deterrence devices, cetaceans find themselves caught in a blinding acoustic fog, unable to detect approaching ships. Human noise pollution renders cetaceans and other marine organisms near deaf, unable to echolocate, communicate, or find prey.[2] Marine traffic accidents are now a primary cause of whale deaths.[3] Although no longer using bullets and bombs, humans are still killing cetaceans by the tens of thousands every year. Could digital technologies provide a solution?

DIGITAL WHALES

The Santa Barbara Channel—through which the world's largest whales, on one of the world's longest migrations, move past some of the busiest ports in the world—is a global nexus for marine roadkill. No better place, then, for developing a Waze for whales. Whale Safe, created by scientists at UC Santa Barbara's Benioff Laboratory, is an AI-powered monitoring system that creates virtual whale lanes, enabling safe passage for cetaceans by preventing ship strikes in real time. The system incorporates five digital technologies: an underwater acoustic monitoring system that automatically detects whale calls; AI algorithms that detect and identify blue, humpback, or fin whales; oceanographic modeling combining satellite and digital buoy data with ocean circulation models and animal tags; whale sighting data reported by citizen scientists, mariners, and whale-watching boats using mobile apps; and locational data from ships' automatic information systems (as mentioned before, a mandatory global system of satellite tracking that enables precise monitoring of ships' locations at all times, in order to prevent collisions).[4]

Combining this information, the system produces a whale presence rating overlaid on a map, similar to a weather report, which is relayed in real time to ship captains, who can decide to slow down or leave the area altogether. The Whale Safe team also tracks ships to see if they are complying with slow-speed zones and publishes public report cards tracking compliance, naming and shaming ships that fail to comply. Scientists are also developing infrared thermal imaging cameras to mount on the bows of ships to detect whales—and whale strikes—in real time. Killing cetaceans used to happen out of sight, but with dashcams mounted on ships, whale sightings will be automatically reported and whale deaths automatically recorded. According to initial studies, Whale Safe has resulted in cutting the number of whale strikes by more than half.[5]

Similar digital whale protection systems have been implemented on the east coast of North America, where aquatic drones now roam the Atlantic, searching for endangered right whales in the Gulf of St. Lawrence, where over 100 million metric tons of cargo are transported each year. The whales' location is pinpointed using digital bioacoustics, their trajectory forecast using AI algorithms trained on datasets of whale movements, and the information conveyed in real time to ships' captains and fishing boats, who face stiff fines of several hundred thousand dollars if they fail to slow down and leave the area.[6] Only a few decades ago, North Atlantic right whales were hunted to the brink of extinction, but today the few hundred remaining whales influence the movements of tens of thousands of ships in a region home to 45 million people. Since the mobile protected areas were implemented, no whale deaths from ship strikes have been reported in the Gulf of St. Lawrence.[7] In stark contrast, along the coast further south, where mobile protected areas have not been implemented, two dozen whale carcasses—most of them the victims of ship strikes—washed ashore in just two months in early 2023; despite this, the recreational fishing industry has continued to lobby against tighter regulations on ships.[8]

What would happen in the future if a whale strike app was hardwired into ships? One possibility, while speculative, is that it would enable whales to control vital aspects of their habitat—like the density and location of ships, and enforceable boundaries for "no-go" zones. Alongside shipping lanes, we would have whale lanes. This is one example of how digital technologies could enable whales (and other nonhumans) to collaborate with humans in managing the environment. Through digital technologies, whales are being enrolled in ocean governance, in stark contrast to the way that humans treated these species only a few decades ago.

Is this an example of what historian Dipesh Chakrabarty calls the extension of ideas of justice to the nonhuman?[9] The potential for nonhumans to guide environmental governance via Digital Earth technologies is reminiscent of a concept coined by philosopher Bruno Latour: Gaia 2.0. Organism-sensor assemblages, connected to an "Internet of Animals," conduct environmental monitoring, shape human decision-making, and invite animals to participate in environmental regulation. Could whale lanes prefigure a more inclusive type of multispecies environmental governance? Marine navigation could begin to systematically include interspecies cooperation, as whales influence and constrain human action by controlling the decisions and movements of ship captains and fishers. We should, however, be careful not overstate the degree of cooperation: the vast power and information asymmetries that exist between humans and other species mean that any form of coregulation would occur on an unequal footing. It is thus inappropriate to ascribe "agency" to the whales in situations where digital monitoring is used to derive data about their movements, in a world in which they have so little control, their ocean territories have been colonized, and their populations have been decimated by industrial whaling.

Admittedly, digital whale lanes would not address all the threats facing marine species, such as chemical pollution and climate change. Nor would they address the underlying causes of ship strikes, such as noise pollution and the rapid increase in container shipping due to globalization and online shopping. More locally made goods, more no-go zones, fewer ships, smaller ships, and slower ships might be more effective solutions. Yet in a political economic climate in which such measures are unlikely, whale lanes may offer one useful conservation tool.

Scientists also point out that these digitally enabled ocean conservation schemes benefit humans as well as the whales. When ships slow down, they not only reduce whale strikes but also release fewer pollutants and emit less carbon dioxide. Moreover, whales' nutrient-rich waste acts like a fertilizer for phytoplankton, which sequester enormous amounts of carbon. Economists at the International Monetary Fund (IMF) have estimated the value of the ecosystem services provided by each individual whale at over $2 million and called for a new global program of economic incentives to return whale populations to preindustrial whaling levels as a "nature-based solution" to climate change.[10]

FOLLOW THE FISH

In an era of rapid global warming–induced changes in the world's oceans, in which many marine species are becoming climate refugees, policymakers are now debating real-time, mobile forms of ocean governance. These systems are relatively low-cost, geographically comprehensive, and agile; as such, they are responsive to real-time adaptation to environmental variability, species mobility, and disturbance dynamics. Digital Earth may be the only workable governance model for an increasingly unpredictable world of environmental hazards and extreme events, and a timely response to the new global commitment—reached at the UN's global biodiversity conference in late 2022—to protect at least 30 percent of Earth's land and water by 2030. Rather than setting aside 30 percent of the world's oceans as marine parks, Digital Earth technology could be used to "follow the fish," enabling the equivalent of marine parks and protected areas to be geographically mobile.

The formal term for this idea is a mobile marine protected area (MMPA).[11] MMPAs are geographically dynamic, with mobile boundaries that change position as endangered species migrate through the ocean. This real-time, mobile form of ocean governance relies on digital hardware that collects data from various sources (e.g., nano-satellites, drones, environmental sensor networks, digital bioacoustics, marine tags, deep sea unmanned underwater vehicles), combined with analytics such as machine learning algorithms and computer vision and ecological informatics techniques. MMPAs are appealing to scientists and regulators because they enable responsive, real-time adaptation to environmental variability, species mobility, and disturbance dynamics.

Why is it so important for protected areas to go mobile? One major reason is climate change, which is driving increased marine dynamism and unpredictability.[12] As the ocean flows, marine organisms follow. Rising global sea surface temperatures, changing ocean currents, and marine heatwaves have led to massive migrations of marine populations. This creates new conflicts between humans and other species. For example, as new zones of the melting Arctic—such as the Bering Strait—open up to shipping, new protected areas are needed to prevent ships from striking whales.[13] But because the ocean's "movescape" is increasingly unpredictable, it is hard to tell where the whales—which follow their food sources—are likely to be. As a result, traditional marine protected areas, with their static boundaries, are becoming increasingly ineffective and irrelevant.

In order for MMPAs to function effectively, they require an enormous amount of data about changing environmental conditions, movements of protected species, and disturbance dynamics.[14] While some data can be derived from satellites, most satellites have insufficient resolution to detect ocean conditions with the necessary precision and accuracy. So MMPAs also use drones and sensors placed on boats, fishing nets, and buoys. Multiple cabled observing networks have also been installed along heavily trafficked coastal areas on several continents.[15] These interconnected systems provide enough data for MMPAs to operate in real time, with comprehensive spatial coverage and both temporal and geographical mobility. But this approach is computationally intensive and requires advanced artificial intelligence as well as digital data collection strategies.

Another reason that regulators are intrigued by the idea of MMPAs is that they promise to enhance enforcement. Traditionally, ocean conservation is reliant on analog techniques, such as human observers on board fishing boats, or logbooks on ships. These analog observational frameworks are hampered by relatively small sample sizes, post hoc reporting, and intimidation of observers. Digital observation enables much larger sample sizes, real-time reporting, and more robust observational mechanisms; no one can bully or harass a camera positioned on a fishing deck, although other forms of gaming the system are, of course, to be anticipated.

PROTECTING THE OLYMPIANS OF THE OCEAN

Do MMPAs actually work? Perhaps the best-known example is the use of MMPAs to protect bluefin tuna, one of the mightiest fish in the ocean, as well as one of the most endangered. Built like torpedoes with retractable fins, an individual tuna can live up to forty years, grow to the length of a whitewater kayak, weigh in at 1,500 pounds, swim as fast as a racehorse, and dive deeper than two Empire State Buildings. Bluefin are also the focus of one of the last great wild fish gold rushes the world may ever see. The fish are in high demand for sport fishing and international fish markets, notably in Asia. An individual bluefin can sell for hundreds of thousands of dollars.

One reason that bluefin are so difficult to protect is that they have one of the longest migrations of any creature on the planet, even longer than humpback whales: one subspecies spawns near the island of Java, and then navigates the Southern Hemisphere's stormy seas to Australia before returning to the Indian Ocean. Some scientists refer to bluefin, which have fascinated scientists since Aristotle's time, as the Olympic athletes of the world's oceans. On these migratory marathons, the fish cross jurisdictions and time zones, confounding attempts at conservation. Making matters more difficult, they change their migration patterns every year in response to changing ocean conditions. Tuna thus turn up as bycatch in fisheries around the world, where they are often discarded in order to save fishing nets or avoid fines.

In the past decade, a new MMPA focused on protecting the critically endangered southern bluefin tuna has been created in the Great Australian Bight, a zone of ocean along the country's southern edge, rimmed by the longest coastal cliff zones in the world.[16] Nicknamed the "Galápagos of Australia," the region is home to a greater number of species than the Great Barrier Reef. The Bight is an ideal habitat for southern bluefin tuna, though studies show the population is now down to 5 percent of preindustrial fishing levels, as rising prices have continued to drive overfishing in one of Australia's most valuable fisheries, despite quotas.[17] Part of the problem is that the endangered southern bluefin tuna species, which migrates from Indonesia to the Tasman Sea, frequents the same parts of the ocean as a year-round tropical eastern tuna longline fishery. Fishers with quotas for eastern tuna inadvertently catch their more endangered cousins, risking high financial penalties.

To combat both overfishing and inadvertent bycatch, a group of scientists and fisheries managers have implemented a mobile management regime based on real-time marine forecasts, also called "nowcasting." The scheme generates real-time predictions of tuna location based on tuna's habitat preferences (which are temperature-dependent). Data gathered through satellite-connected fish tags, which track the fish in real time, is combined with temperature data (satellite sea surface temperature data and vertical temperature data from an oceanographic model), plus habitat preference modeling, to create real-time predictive maps. The maps, which are updated daily, locate hotspots for southern bluefin tuna to within a few meters, and predict seasonally adjusted southern bluefin tuna locations sixty days into the future, allowing fishers to easily visualize the distribution of "no-go" fishing zones over the coming months and plan their fishing trips accordingly.

The size and borders of no-fishing zones designed to protect the tuna are updated daily, fluctuating in response to predicted fish location. Fishers can adapt their behavior on a monthly, weekly, or even daily basis. The dynamic spatial zoning enabled by digital and often automated methods has proven to be more accurate than the subjective, manual approach previously used by fisheries managers.[18] In 2015, the Australian Fisheries Management Authority implemented another layer of oversight: video monitoring of every set in the fleet, enabling verification of each fish captured against the permitted quotas for each boat. The result: lower bycatch and more systematic quota management. The system is working to protect the tuna, without undue disruptions to fishing. Based on these results, scientists have argued that MMPAs can achieve significantly better biodiversity conservation outcomes than conventional, static protected areas, particularly as climate change accelerates shifts in species ranges.[19] Dynamic fisheries closures, a specific type of MMPA that temporarily prohibit fishing in some areas, may also reduce bycatch much more effectively than static marine protected areas. In an analysis of fifteen fisheries around the globe, scientists found that bycatch would be reduced up to 57 percent using dynamic marine protected areas, but by only 16 percent if the protected areas were geographically fixed.[20]

COULD MOBILE PROTECTED AREAS GO GLOBAL?

The protected areas that now surround the Pacific bluefin tuna may be a harbinger of things to come, as a growing number of scientists believe that this type of dynamic management will become a cornerstone of marine conservation in a climate change–challenged world.[21] Of course, humans have tracked animal movements for millennia for survival and hunting, as well as for managing and protecting wildlife populations. The difference today is the unprecedented degree of digital surveillance of which humans are now capable. And new types of Digital Earth technologies, such as bioacoustics and ecoacoustics, enable the detection of animals by hearing them, without even seeing them. With these recent advances, the digital technology necessary for mobile marine protected areas now exists in some regions of the world's oceans.

However, as with whale lanes, MMPAs are not a panacea. They do not address the full scope of threats to marine creatures, such as chemical and plastic pollution. Another concern pertains to the use of smart oceans technologies for economic and military purposes. For example, DARPA, the US military research agency whose research gave rise to the internet, is one of the world's leaders in smart oceans, which it calls its Ocean of Things program. Digital undersea technologies are also being used in deep sea mining exploration, which uses seismic blasting with well-documented negative effects on marine species. Digital infrastructure for marine conservation is thus, at times, associated with processes of militarization and digital colonization of the world's oceans for resource extraction.[22]

Issues of distributional justice are also a concern. To date, MMPAs have been largely implemented by wealthy countries; cross-subsidy regimes are possible but have not yet been developed within global marine governance regimes. Moreover, MMPAs have been focused on a small number of "charismatic" species, such as whales or high-value species such as tuna, rather than species such as herring or sardines, which have high ecological importance but lower profits for fishers. MMPAs could risk distorting marine conservation regimes and reinforcing existing patterns of justice, unless this issue is thoughtfully addressed in an integrated manner that prioritizes the needs of poor and vulnerable communities.

MMPAs are also technically challenging to implement. Complex ecosystems are characterized by distinct and often variable spatial and temporal scales. At times, these scales may be overlaid and interact. For example, climate and ocean circulation patterns fluctuate over years and decades, whereas fish location fluctuates on a daily basis. In some cases, scientists do not yet have a sufficiently developed understanding of ecosystems to determine which variables must be monitored, and on which time scales. In these cases, digital technologies are helping to deepen our understanding of these ecosystems, but these systems are not yet ready for mobile governance, as "now-casting" is not always possible. The best option in these cases is to use digital monitoring to track marine organisms as they retreat, disperse, and flee in response to climate change, and provide a more responsive approach to spatial relocation of protected areas to new habitats. The establishment of MMPAs on a global scale will also be hampered by gaps in global ocean governance. On the high seas, individual nations do not have jurisdiction. Existing international mechanisms to govern the high seas are slow to change, particularly because the Law of the Sea—which originally focused on pollution prevention and fishing—was not designed to address issues such as biodiversity and climate change.[23]

The passage of a new UN High Seas Treaty in 2023 is thus a hopeful sign, as it opens the door to creating new Marine Protected Areas. Researchers have already identified biodiversity "hot spots," which are the top priorities for protection.[24] A team of researchers at the Benioff Ocean Science Laboratory at UC Santa Barbara, for example, conducted an analysis that integrated fifty-five global data layers on biodiversity (including over 12,000 species), ecosystem features, productivity, and fishing in order to systematically analyze regions that should be prioritized for protection, such as the Costa Rica Thermal Dome, a nutrient-rich region frequented by endangered blue whales and leatherback sea turtles; the Hawaii-Emperor Seamount Chain, a vast range of largely underwater volcanoes that are home to some of the world's oldest living corals; and the Indian Ocean's Mascarene Plateau, one of the largest shallow areas in the global ocean, home to the largest seagrass meadows in the world.

When it comes to implementation, several questions remain.[25] One question is what to do with existing marine protected areas, which have static boundaries. An interconnected tapestry of static and dynamic marine protected areas is likely to be the optimal approach. Dynamic management can provide more effective protection of mobile species, but static protected areas are still important and relevant, particularly for fish nurseries and areas of high biodiversity. So static and dynamic management frameworks will likely coexist.

Another question is how mobile protected areas will be monitored and enforced. Governing the open ocean requires data-intensive technologies, and developing countries have emphasized the critical importance of embedding commitments to data access, pushing for associated capacity building and technology transfer for data collection, and support for enforcement.[26] Although satellite data can identify illegal fishing, local enforcement agencies—often underresourced—are still required to apprehend those breaking the law.[27] The new UN treaty on the high seas contains provisions to ensure that digital data is shared with lower-income countries.[28]

Mobile protected areas also raise significant questions regarding data privacy. Digital environmental surveillance technologies may be misused for surveillance of humans or may simply accidentally monitor humans, a phenomenon known as "human bycatch." Digital ocean monitoring could enable better information access for scientists and regulators, but could also enable digitally savvy criminals to accelerate illegal fishing. Staying one step ahead of illegal fishers now requires a commitment to digital resources that few countries can afford. In light of these developments, the efforts of organizations like Global Fishing Watch and innovators like Dyhia Belhabib (discussed in chapter 2) will become even more important.

MOBILE, MULTISPECIES ENVIRONMENTAL GOVERNANCE

Mobile marine protected areas challenge two core assumptions of contemporary politics: political boundaries are fixed and static, and nations have the right to claim territories within certain boundaries as their exclusive possessions. Nearly every square mile of land on Earth is owned by *Homo sapiens*. There are only a few exceptions, usually land in places uninhabitable for humans: Antarctica or Bir Tawil, a tiny sliver of desert between Egypt and Sudan. These zones are known as *terra nullius*: a Latin expression meaning "no one's land," interpreted to mean a zone of land that no state claims or occupies. Colonialism and the forging of nations have, over the centuries, created a political condition that seems normal to us, but is in fact a historical anomaly. National sovereignty is predicated on monopoly control over land, backed up by force.

In the past century, successive reforms to international law have extended national control from the land to the sea. At first, control reached only a few miles beyond the coastline; now it goes as far as 200 miles. But the vast majority of the world's oceans, the high seas, remain ungoverned. These areas of the globe were once beyond reach and largely unseen, but digital technologies extend the domain of state sovereignty via planetary-scale computation.

But there is also a deeper change occurring. Digital monitoring of the world's oceans at a global scale is an example of planetary computation. The availability of marine data in real time, anywhere around the world, creates a new sense of responsibility and possibility. The new UN treaty on the high seas is an example of planetary governance that mirrors—and, indeed, is made possible by—planetary computation.

The implementation of mobile protected areas under this new planetary ocean governance framework will challenge a core assumption in environmental management and geopolitics: boundaries must be fixed, static, unmoving in time and space. Mobile management of biodiversity is geographically agile. As boundaries become fluid, the principle of environmental stewardship becomes more actionable. In the longer term, it is possible to conceive of comprehensive digitization of monitoring and management of protected areas on the high seas. AI-supported decision systems could incorporate a large range of complex interactions between natural systems (e.g., climate-ocean coupling, shifting food webs) and social systems (e.g., marine pollution, overfishing). Humans once drew boundaries on maps; often, colonial overlords did so from far away. Is an automated AI decision-support system a liberating innovation, or a digital overlord?

We will soon face similar questions on land. Over a century ago, the creation of national parks became enshrined as a cornerstone of conservation. At the time that parks were created, it was recognized that their boundaries are porous and often insufficient to protect highly mobile species. Nonetheless, parks and protected areas became the jewels in the crown of the twentieth-century environmental movement: iconic and much-loved symbols. However, static protected areas may no longer be sufficient to protect species and habitats in an era of climate change. As our climate shifts, so too does human land use, habitats, species ranges, and phenology (the timing of recurring biological events, like flowering and animal migration, which is strongly influenced by climate). But what if parks and protected areas could move in both space and time?

This is one example of how Digital Earth monitoring creates multispecies networks, enrolling what Achille Mbembe refers to as "le vivant" (the living, enlivened, lively world) into environmental governance.[29] Digital technologies may allow nonhumans to participate as active subjects in environmental management. Planetary computation, in other words, is not merely a set of tools for monitoring and manipulating the planet but also a potential means of extending political voice to nonhumans. Digital Earth networks are thus not merely extensions of the old engineering mantra of "command and control." Instead, they offer us a new paradigm: "communicate and cooperate," which extends a form of voice to nonhumans, who become active subjects co-participating in environmental regulation, rather than passive objects. The environmental becomes inescapably political, but the political is not solely human.

A mobile marine protected area enables different organisms to express preferences, share space, and collaborate to sustain habitability of ecosystems and communities. Might this be extended to other species, and generalized? This depends on whether nonhumans can indeed communicate their preferences via signals—biochemical, vocal, vibrational, gestural—that can be detected by digital monitoring systems. If these signals can be detected, and meaningfully incorporated into decision-making systems, this implies the possibility of a very different future for environmental governance: one in which nonhumans can participate.

A DIGITAL PARLIAMENT OF EARTHLINGS

In *We Have Never Been Modern*, Bruno Latour argues that humanity has entered a new era in which scientific innovation enables the proliferation of hybrid technologies. These hybrids have a fundamental effect on human politics because of the challenge they offer to a long-standing binary distinction in Western thought: humans versus nonhumans.[30] Hybrid technologies, Latour argues, create a new kind of hybrid politics; Isabelle Stengers refers to this new political dispensation as "cosmopolitics." Latour and Stengers argue that technology, science, and politics coexist and mutually shape one another; as technological innovation changes our relationship with the nonhuman world, so too our decision-making processes change. In subsequent books, Latour argued that nonhumans should be part of our democratic decision-making processes, both because they are affected by our actions and because they have what social scientists term "agency": the ability to make free choices and act independently. As Latour argued in *Facing Gaia* in 2017, nature never played the seemingly passive role to which it was assigned by Western science.

Latour articulates the need to invent new political systems that account for the fact that human lives are intertwined with nonhumans, with ecosystems, and with the planet itself. He invokes the ideas of French philosopher Michel Serres in arguing that humanity needs a new type of "natural contract," based on ideas of symbiosis.[31] Whereas humans currently act like parasites, avidly depleting the resources of our host without seeming to realize that we are undermining our own health and lives in the long run, we should act more like symbionts, creating mutually beneficial relationships with nature. Humanity needs, he argued, a new type of democracy, which he called a Parliament of Things.

No specific explanation of *how* the Parliament of Things would work is offered by Latour, although he offers preliminary thoughts in *Politics of Nature*.[32] Rather, he leaves the reader with an intriguing but incomplete agenda: moving beyond technologies of control to technologies of negotiation. By this, he means technologies that enable us to listen and hear, strategies for becoming more aware of and sensitive to the "ways of being" of nonhumans, to their reactions, behaviors, and relationships. Scientists and innovators could understand their role as diplomats or perhaps as shamans. Rather than—or in addition to—acting as experts or engineers, they could mediate between humans and nonhumans. How precisely such mediation is supposed to occur is not specified; Latour's Parliament of Things is an abstract thought experiment.

As I argue, Digital Earth technologies offer a way to instantiate the Parliament of Things. Digital technologies convey real-time information about other living organisms—their locations, their preferences, the conditions they require to lead safe, healthy lives. As the example offered at the outset of this chapter demonstrates, this data can be incorporated into digitally enabled environmental regulation systems that constrain human behavior in order to protect the rights and freedom of nonhumans. Building on Latour, I suggest that these systems be called Parliaments of Earthlings: digital systems that enable the exchange of information between species, and a means for a collective of nonhumans to guide and perhaps constrain human behavior and choices.

What is the difference between an Internet of Earthlings and a Parliament of Earthlings?[33] The former focuses on what I termed in chapter 1 the "eco-tech stack": the hardware and software, wetware and dryware, that connect different organisms to digital monitoring systems; from an engineering perspective, this is a simple extension of the Internet of Things. The latter is a political concept, focused on questions of collective action, power, agency, justice, and equity. A functioning parliament depends on mutually intelligible dialogue: a mechanism for exchanging information, based on concepts that have stable, shared definitions. This dialogue is enabled by Digital Earth technologies, which create a panopticon of digital environmental surveillance: sensor networks and planetary computers that continuously monitor and analyze Earth and its inhabitants, the living and the nonliving, from the depths below the surface to the upper atmosphere.

The word "parliament" derives from the French *parler* (to speak), which in turn derives from Latin and Greek words for public expression and comparison (to set side by side). Medieval references to discussions with rules (termed *parlement* or *parlamentum*) gradually grew to refer to the gathering itself. A parliament is thus not necessarily fully democratic, although the term has come to be associated with liberal democracies. Indeed, the birth of Western democracies was closely associated with new ways of sharing information and collective spaces for dialogue among those of different backgrounds and identities.

This Parliament of Earthlings is democratic, in the etymological sense of the term: it enables rule by both human and nonhuman individuals. But it depends on mechanisms different from those of conventional human parliaments. Voting, in the sense of periodic elections and ballot boxes, is not the primary mechanism by which the Parliament of Earthlings will function. Rather, the continuous sharing of digital information about nonhumans, which influences human action, is analogous to political voice. The question of political voice might mean, simply, the ability to express preferences that are interpreted by humans and influence human conduct. But this presupposes the point that nonhumans can communicate in some fashion, and that their communicative utterances contain meaningful information. Can they?

CAN EARTHLINGS SPEAK?

Humans might choose to listen to nonhumans, but do they have anything meaningful to say? Here, too, Digital Earth technologies offer insights.[34] As I explored in my previous book *The Sounds of Life*, digital bio- and eco-acoustic networks have been deployed around the planet, from the Arctic to the Amazon, recording and decoding nonhuman sounds—many of which occur beyond human hearing range in the high-frequency ultrasound or low-frequency infrasound. The proliferation of these digital listening systems reveals that much more meaningful information is encoded in acoustic communication within and between species than humans suspected.

The Sounds of Life makes three arguments. First, a much broader range of species than previously understood can detect, respond, and make sound. Many species that scientists once thought to be mute, or relatively vocally inactive, actually make sound. To give just one example, researchers have recorded hundreds of different sounds made by over fifty fish and turtle species—once thought to be voiceless—that reveal complex coordination behaviors, evidence of parental care, and a remarkable ability of embryos of at least one species to time the moment of their collective birth through vocal communication. Peacocks emit loud, low infrasound during their mating dances. Elephants use similar frequencies to communicate across long distances, seemingly telepathically. On a coral reef, creatures share information acoustically; coral and fish larvae, creatures without ears, can sense the sound of healthy reefs, and even orient themselves to the sound of their home reefs. Nearly every species to which scientists have listened makes some form of sound. Nonhumans were once believed to be largely deaf and mute, but now we are realizing that in nature, silence is an illusion—although much of this communication occurs beyond human hearing range, and thus is audible only through the use of digital technologies. The world is resonant with nature's sounds, which human ears cannot detect—but our computers can.

The second main argument of the book is that nonhumans are not merely making and responding to sound; they are communicating with one another, and in much more complex ways than Western science has previously understood. Elephants, for example, have specific signals for different threats such as honeybees and humans, and their vocalizations even distinguish between humans from different tribes; researchers are now building an elephant dictionary with thousands of sounds. Honeybees also have hundreds of distinct sounds; although we have deciphered only a few, we know that there are specific sounds in honeybee language—which is spatial and vibrational as well as acoustic—with specific meanings, such as a "stop" signal and a begging signal.

Digital listening is also revealing that interspecies communication is much more widespread than scientists previously understood. Moths can detect and even jam the sonar of bats. When buzzing bees approach flowers, the flowers respond by flooding themselves with nectar within minutes. Plants can detect the sound of specific insect predators and distinguish threatening from nonthreatening insects with astonishing precision. Corn, tomato, and tobacco plants emit high-pitched ultrasound that we can't hear, but insects likely can; in one experiment, researchers trained an AI algorithm on the distinct sounds emitted by healthy, dehydrated, and wounded plants—and the algorithm was soon able to diagnose the plants' condition, simply by listening. Could other creatures be listening to the plants, and detecting their state of health?

The third argument of the book is that, by using AI to decode patterns in nonhuman sound, we can gain new insights into how nonhumans communicate, while also enabling rudimentary attempts at two-way communication, mediated by robotics. Digital technologies, in short, offer new ways of engaging with nonhuman agency and voice. This is by no means novel (Indigenous traditions offer powerful ways of nonhuman listening) nor neutral (digital technologies can be misused and abused). But with caveats and safeguards, digital bioacoustics offers humanity a powerful new window into the nonhuman world.

The nonhuman world is resonant with mysterious sounds, which scientists are only beginning to decode. And sound is only one modality of nonhuman communication; other species use many mechanisms—from the gestural to the biochemical to the electrostatic to the vibrational—to communicate information and even emotions. Vibrations along a spider's web convey information; the spider sees the world through its web, which allows it to distinguish a gust of wind from the trembling of a fly. Digital technologies can detect these multimodal forms of information, whether from forests or honeybees. The internet is just one of many systems that enable living species to convey information.

As humans deploy digital technologies to enhance our ability to monitor and decode this information, we create the potential for a new type of environmental governance, and a new type of multispecies politics. If communication is ubiquitous in nature, and is not unique to humans, then nonhumans can be said to possess a political "voice." Indeed, nonhumans already exercise voice, in other words; but modern humans have been hard of hearing. In the future, digital technologies may not enable nonhumans to vote, but it may be technically feasible for them to exercise political voice, as the complex information that they communicate can now be transduced, transcribed, and translated by digital technologies. The Parliament of Earthlings would thus be partly mediated by digital bioacoustics, which allow humans to better grasp the sentience and complex communication capacities of a broad range of species, and perhaps even help us break the barrier of interspecies communication. Nonhumans have much to say to us, and can no longer be dismissed as mute and or ignored as unknowable.

Complex communication is ubiquitous in nature, and thus many nonhumans could be said to possess a form of political "voice." Digital Earth technologies offer new ways of listening to nonhuman preferences. Planetary computation enables multispecies conversations, in which digital technologies sense how nonhumans convey information; if this information is used to influence human action, could this be a form of political voice? How might nonhuman preferences be incorporated into our decision-making frameworks, into new forms of earthly politics? To begin formulating an answer, we'll have to listen more closely to our nonhuman kin.

We will also have to rethink what we understand by "political voice." It may mean co-participation in environmental regulation, expressing preferences and choices, and influencing human behavior, but it is not narrowly confined to voting. The principles underlying the granting of political voice, the legitimacy to speak, may also evolve and become rooted in geographical territory. In most current nations, political rights are tied to citizenship; as we pay greater attention to nonhumans, we may come to see dwelling and being in place (or, as sociologist Vanessa Watts puts it, place/thought/being) as the basis for political representation. This implies that principles such as reciprocity, regeneration, resilience, and symbiotic autonomy might supplement (or even supplant) principles such as freedom, liberty, and equality. Yet the history of the internet demonstrates that such principles are difficult to defend in a digital world. Decades ago, the French philosopher Michel Foucault offered a critique of the practice of biopolitics: systems that administer and surveil, and hence inculcate the internalization of control by power. Many digital technologies in this book *could be* biopolitical paradigms of control, aligned with surveillance capitalism. But digital technologies could equally be mobilized to expand progressive possibilities, extending the political franchise to nonhumans and expanding our socionatural commons. Digital Earth technologies simultaneously offer regressive and progressive possibilities for humans and nonhumans alike.

HONEYBEE DEMOCRACY

We are left with one nagging question. Latour's suggestion—that we can understand what nonhumans want merely by listening to or negotiating with them—seems problematic. Our knowledge of the world is always mediated, whether by language, ideology, power relations, social structures, or (mis)representations of others. This challenge already makes political relationships among humans difficult enough. For this reason, even if extending democracy to nonhumans was desirable, critics have argued that it seems infeasible—at least as democracy is currently conceived. It is likely, then, that extending the Parliament of Earthlings to nonhumans will also change the way democracy is practiced. But how could digital technologies enable nonhumans to truly participate in *deliberative* democracy? Perhaps we should look for answers that nonhumans have already devised to this question. What if we were to ask honeybees?

When scientists first tagged honeybees with RFID chips, protecting the bees from harm was their primary goal. Many similar studies around the world have had the same agenda. Microchips have now been used to track everything from dragonflies and crickets to rhinos. But once scientists started to wire up honeybees, they began surprising us with other insights. Decoding the diversity, density, and interactions in honeybee swarm patterns, for example, has led researchers to describe the swarm as a cognitive entity that rivals the human brain in its decision-making power. Cornell researcher Thomas Seeley has demonstrated that when searching for a new site for a hive, honeybee swarms democratically debate the options, and gradually reach a consensus before collectively flying the route to the new, preferred home. Indeed, when the collective is arriving at a decision, the interactions of individual bees bear a remarkable resemblance, Seely argued, to the interactions between our individual neurons, and exhibit sophisticated forms of democratic decision-making, including collective fact-finding, vigorous debate, consensus-building, and quorum.[35] The bee swarm, in other words, is a democratic decision-making body in motion.

Seeley's findings were made possible through the use of computer vision. Digitalizing the swarm enables scientists to document honeybee behaviors, revealing findings that would otherwise remain inaccessible. How could environmental governance frameworks not only absorb data from hives but also incorporate the principles by which honeybees make decisions? In other words, should the ways in which honeybees make decisions inform our own decision-making practices when what we do is likely to impact them? This might sound outlandish but is perhaps not greatly different from other legal principles, such as respect for sovereignty and the responsibility to protect. Animal rights scholars have long argued that these rights should be extended to nonhumans. Digital technologies provide a mechanism by which this might occur on the animals' own terms. Perhaps we might create different Parliaments of Earthlings in different places. Could honeybee decision-making eventually influence how humans make decisions about, say, agriculture in the Great Plains, just as whales determine the paths of ships in the North Atlantic?

Such speculations seem tempting, in part, because they play into an archetypal revenge fantasy: nature taking back the Earth. In reality, mobile marine protected areas or honeybee monitoring systems do not exist on a level playing field; the recognition of nonhuman "voice" does not automatically imply that humans will cede control to other species. Data as derived from digital technologies may lead to closer monitoring, but there is no guarantee that this will lead to more autonomy for nonhumans, or habitability and safety of our planet. Digital technologies enable voices to be monitored, but do not require that they be included; it is a fine line between surveillance and political dialogue.

How, then, might we mobilize digital technologies to enable new types of reciprocal relationships with nonhumans that share power and resources more equitably, that are attentive to the varied experiences of environmental degradation occurring in different places and to different species and individuals? Do nonhumans need some kind of legal standing, or some type of rights? In the next chapter, I explore how digital technologies may enable legal personhood and property rights for nonhumans.

PARABLE OF TERRA0

From an airport in Dubai, a private jet takes off.

On the AirCarbon Exchange, a trader buys a carbon offset. She chooses a blockchain-verified tree-planting credit. Her payment is deposited in a bank account in Singapore.

An algorithm communicates instructions to its tree-planting drones in a forest in India. Guided by a forest sensor network, calibrated with climate data, the drones optimize the planting sites for the new seedlings. The trees will be monitored by soil sensors and cataloged by satellite as they grow, the newest members of a self-sovereign forest named terra0.

Once, humans came and named the trees, marking them with sacred scars. Now, each seedling has a unique bar code, an unchangeable digital label.

The profit from the carbon offset will be used to buy blockchain-verified fractional shares, owned by the forest, in land on terra0's eastern border. The algorithm estimates that it will take several thousand years to buy back and replant its traditional territory. But trees live a long time, and have many children. And on the blockchain, no one knows you're a forest.

7 ON THE BLOCKCHAIN, NO ONE KNOWS YOU'RE A FOREST

In the previous chapter, I argued that Digital Earth technologies create the possibility for multispecies environmental regulation. By this, I mean a form of environmental regulation in which human action is constrained, shaped, and guided by the preferences and information shared by nonhumans via digital networks. In describing this concept, which I termed a "Parliament of Earthlings," I noted one major shortcoming: the power imbalances that exist between humans and other species. In the absence of other legal foundations of democratic participation—such as legal personhood and property rights—the use of digital technologies to monitor Earth would be likely, I suggested, to degenerate into eco-surveillance. But I briefly hinted at the idea that digital technologies could offer a mechanism whereby certain types of rights could be enacted for nonhumans. This latter argument is the focus of the current chapter.

Digital technologies add a complicated twist to the long-standing debate over environmental rights, also known as "rights of nature." The rights of nature have been debated for decades, and although not yet mainstream worldwide, they have gained legal traction in some countries. Cities, states, and countries have enacted legislation that grants rights to living beings such as animals and forests, and nonliving entities such as rivers and lakes. In 2006, the small borough of Tamaqua, Pennsylvania, banned toxic sludge disposal as a violation of the rights of nature; since then, rights of nature laws have been implemented by dozens of communities in ten states across the United States, ranging from small cities like Broadview Heights, Ohio, to large cities like Pittsburgh. Ecuador's constitution recognized the rights of nature in 2008, followed by laws protecting the rights of nature in Bolivia (2010), Colombia (2016), New Zealand (2017), and India (2017). Many of these laws and declarations recognize nonhumans as entities with some form of nonhuman personhood.

However, despite being increasingly recognized, such rights have proven to be difficult enforce, and many view them as largely symbolic. One thorny issue is a question that concerns the scope of these rights: Should nonhumans merely be granted the right to exist? Or should they be granted more expansive rights, such as the right to a life free of pollution? Should rights be granted to individual organisms or to entire ecosystems? And what are the consequences of recognizing nonhumans as legal individuals that have the right to exist, yet are incapable of holding other rights? In this chapter, I explore how digital technologies might advance these debates. After briefly summarizing the history of the rights of nature debate, I focus on the use of digital ledger technology—often referred to as blockchain—to create nonhuman property rights.

THE RIGHTS OF NATURE

Laws that grant property rights to nonhumans have many historical precedents. In France, for example, easements are sometimes attributed to nonhuman entities such as villages (which might own a nearby park or grazing area in common), historical buildings, or sites of environmental importance such as a river delta wetland, which provides a refuge for migrating birds.[1] While not having the status of legal persons, nonhumans may still exercise property rights. Certain entities—places, buildings, and even monuments—have the status of a property owner. When dwelling or passing through land that is subject to such an easement, something profound happens to human beings: a human is no longer a subject/person who dominates places as objects/things. Rather, a human being becomes part of a community, which includes nonhumans. Humans might have the equivalent of a residency permit, but they do not own the land on which they reside; the land owns itself, and is also recognized as the provider of beneficial services—such as fresh water or a sheltering place—for other species.

Another starting point for the debate over environmental rights is the Five Freedoms framework, which was developed in the 1960s to address the rights of animals and has been accepted into law (in modified forms) in various countries: freedom from hunger, thirst, and malnutrition; freedom from fear and distress; freedom from physical discomfort; freedom from pain, injury, and disease; and freedom to express normal patterns of behavior.[2] Beyond this, advocates argue for a broader set of rights and freedoms that enable, as animal rights advocates put it, "a life worth living."

Nonhuman property rights are another concept that has recently gained traction. Should nonhumans be able to hold sovereign property rights to themselves? And does this require that we grant nonhumans the status of legal persons? In some Western countries, both humans and their corporations can own property and are recognized as legal persons. The debate over legal personhood for corporations has a long and heated history. The question of whether corporations should be recognized as legal persons (first recognized by the US Supreme Court in 1886) is a particularly contentious point. Legal scholars and activists have argued that the creation of legal personhood for corporations creates a precedent for granting the status of legal persons to nonhumans. This idea has recently been tested in a series of high-profile court cases that have argued that primates and other nonhumans are legal persons. A number of countries have passed laws granting legal standing to animals (Argentina, Denmark, Pakistan) and to environmental entities such as rivers (Bolivia, India, New Zealand).

A similar idea was put forward in 2015, when representatives acting on behalf of a creek in Pennsylvania moved to intervene in a state court lawsuit brought by an energy company against the local municipality, which had enacted a law banning the drilling of oil and gas wastewater wells; the motion was brought by a local citizen's group, which named Little Mahoning Creek as the intervener.[3] This was not the first time a river went to court: lawsuits have been filed on behalf of rivers, marshes, brooks, beaches, trees, turtles, birds, and—perhaps most famously—whales, when the global cetacean community took President George W. Bush and Secretary of Defense Donald Rumsfeld to court. The *Cetacean Community v. Bush* decision of the US Court of Appeals for the Ninth Circuit in 2004 definitively concluded that animals do not have legal standing, much less legal personhood, dashing the hopes of animal rights advocates in the United States.

Internationally, a growing number of governments have recognized nonhuman rights. The Denmark Animal Welfare Act (2020), for example, declares that animals are sentient beings with inherent worth and behavioral needs that must be respected. Perhaps even more significantly, in 2008, a new constitution was passed in Ecuador with an entire chapter devoted to granting inalienable rights to nature to exist, persist, and be restored where damaged. Based on Indigenous Quechua and Aymara concepts of Pachamama (the Andean Earth goddess), the new constitution's preamble says that every woman, man, and sovereign person "hereby decide[s] to build a new form of public coexistence, in diversity and in harmony with nature." The constitution, which was put to a general vote of Ecuador's human citizens, passed with two-thirds in favor, thus granting the country's nonhuman residents—monkeys and snakes, turtles and orchids—constitutional rights.

Four years later, Bolivia passed its Mother Earth Law (Ley de Derechos de la Madre Tierra), the first modern example of a law that gives legal personhood to nonhumans, inspired (like Ecuador) by the Indigenous communities of the Andes. Bolivian citizens can now sue individuals, groups, or companies on behalf of Mother Earth. In 2017, Colombia's Constitutional Court granted legal personhood to the Río Atrato in the Chocó biodiversity hotspot, while the Ganges, Yamuna, and Narmada Rivers were recognized as living entities with legal rights by the courts and legislators in India.[4] That same year, after decades of court cases and negotiations with the Māori, New Zealand's parliament passed a law recognizing the Whanganui River as a legal entity with the "rights, powers, duties, and liabilities of a legal person."[5]

Is legal personhood an effective vehicle for the rights of nonhumans? Some critics warn that extending rights—a human invention—to nonhumans may backfire, particularly where the concept of legal personhood is inherited from a colonial legal tradition. Some Indigenous legal systems, as the Anishinaabe legal scholar John Borrows has argued, recognize nonhumans as kin, a relational concept that is distinct from the individualistic notion of personhood in the common law legal tradition.[6] Legal pluralism—the coexistence of colonial settler law and Indigenous law—has indeed gained traction in recent years in some countries (including Bolivia, Canada, and Ecuador). Anishinaabe scholar Aimée Craft has explored how Indigenous practices of treaty-making can be extended to nonhumans, offering reconciliation and reciprocity rather than re-colonization. Craft argues for a "braiding" of international, domestic, and Indigenous laws, and thus for legal pluralism.[7] As she explains, there are concepts other than rights within non-Western legal traditions that might be better suited to the recognition of the autonomy and sovereignty of other species: for example, treaties can be passed between human and nonhuman communities.

Critics also argue that these proposals will accomplish little in practice. Existing legal systems have not, in general, proven to be effective at enforcing environmental laws. Will Ecuador's estimated 2,000 species at risk—the highest number of any country in the world—be better protected by the new constitution, given the country's ongoing dependence on oil revenues and poor track record of preventing pollution from oil companies in the Amazon? In fact, some evidence points in that direction. In a 2022 court decision, Ecuador's Constitutional Court handed down a landmark decision that blocked a mining project in the protected Los Cedros cloud forest; citing the high biodiversity in the forest, the court found that mining would violate the rights of nature. As a result, in the future mining companies will need to undertake much more stringent studies before beginning projects. They will also be required to engage in precautionary and restrictive measures to ensure that their projects do not harm endangered species or the ecosystems on which they depend.[8]

These lawsuits, while inspiring, nevertheless have significant drawbacks. First, they are only a post hoc solution: the environmental damage must be done before anyone can go to court, and litigation is both time-consuming and enormously expensive. Moreover, they depend on humans litigating or acting on behalf of nonhumans. The underlying problem is this: even though nonhumans are now legal persons, they are—given our contemporary political and economic norms—the most powerless people on the planet. They are denizens, but not citizens. To address this issue, some conservationists are now proposing that we use a new innovation—digital ledger technology—in an attempt to enable nonhumans to both have and wield property rights.

CRYPTO-CONSERVATION

Imagine a digital database that immutably stores and tracks the history of every transaction on a network. The list of transactions—known as a ledger—is distributed to all network users and stored simultaneously on multiple, interconnected computers; no central authority is needed to store or access the information. Each computer—also called a node—validates each piece of data; this creates a record of transactions, while establishing a distributed, verifiable consensus on the validity of the dataset across the entire network. While new transactions can be added, prior transactions cannot be edited. The digital ledger is thus both distributed and immutable. And because the network is decentralized, with no central authority, it is referred to as a peer-to-peer network.

Perhaps the best-known example of distributed ledger technology is blockchain. Data about successive transactions is stored in a "block," sequenced in a "chain"; information about the transaction is encrypted via a "hash" that also encodes information about previous transactions, making the blockchain network supposedly tamper-proof. Not all distributed ledgers are blockchain, the main difference being that some distributed ledgers are owned and controlled by a single entity. Moreover, not all distributed ledgers use encryption, nor do they necessarily group cryptographic blocks in a chain. However, in the popular press, the term "blockchain" is often used—although imprecisely—to refer to distributed ledger technology more broadly.

In the past decade, the environmental conservation community has experimented with different types of distributed ledger technologies. Although the best-known application of digital ledger technology is cryptocurrency, a wide range of data can be stored in the ledger, including data relevant to environmental sustainability.[9] Blockchain applications have been proposed for a variety of environmental applications, including tracing provenance in supply chains, thereby rendering them more transparent; financing climate change mitigation through using blockchain to sell carbon credits; and enabling direct-to-affected-communities philanthropy.[10] As an example of the latter, the NGO GainForest uses smart contracts to enable the transfer of donations directly to Indigenous communities to protect their traditional lands; the funds are released once specific criteria—such as determining the proportion of land that retains forest cover, automatically using machine learning algorithms to assess satellite imagery—are met.

Proponents argue that this use of blockchain enables a direct link between philanthropists and communities, while enabling transparency about where funds flow, providing checks against fraud and money laundering, and confirmation of whether funding goals are met. Major environmental organizations, including the World Wildlife Fund and the Global Environmental Forum, have praised the potential of blockchain because of its ability to incorporate "smart contracts," which are programmed to automatically execute commands when certain conditions are met, without the need for validation by intermediaries or auditors. Advocates have thus argued that digital ledger technology could reduce or even eliminate corruption and malfeasance.

Critics, however, argue that "crypto-giving" is little more than surveillance philanthropy, which enables donors to more tightly control the beneficiaries of their largesse, limiting their autonomy. The hefty fees sometimes charged by intermediary platforms like AidChain, Promise, and Tokens for Humanity also raise the costs of donor transactions in a manner that may be prohibitive for smaller charities. Critics also point out that it is misleading to portray blockchain as a trust-based technology without intermediaries; rather, those who control digital technologies—writing code, launching satellites, creating algorithms to analyze data, designing smart contracts—become a new intermediary, and one that is arguably less transparent, insofar as the underlying technology is opaque to most users. Moreover, blockchain is often closely connected to a particular form of environmental conservation that advances the quantification and commodification of nature through strategies such as payments for ecosystem services. Blockchain philanthropy, in other words, is not neutral; rather, it is an expression of a particular socioeconomic worldview, closely linked to platform capitalism. But could there be other uses of digital ledger technology that avoid these pitfalls, and which focus not on conventional donor economies but rather on creating more autonomy for nonhumans?

TERRA0

In recent years, blockchain has been touted by libertarians advancing notions of freedom, autonomy, and sovereignty beyond the nation-state. Much of this energy is directed at subverting government-controlled monetary systems. As Larry Lohmann puts it, blockchain and smart contracts promise a world "of order without law. Of finance without banks. Of regulation without regulators. Of trust without governments . . . a world in which you could safely own and trade private property without having to involve other human beings at all."[11] The ongoing volatility in the cryptocurrency market, and the spectacular collapse of the cryptocurrency trading platform FTX, are stark reminders of the flaws in libertarian arguments and importance of regulatory safeguards for global financial systems.

Blockchain technology is thus an ambivalent tool for environmental governance. Yet digital ledger technology is not synonymous with blockchain and may offer other, more progressive strategies for enhancing environmental monitoring and nonhuman autonomy. Consider the following example. Created in 2016, terra0 is a speculative innovation—a self-owned, autonomous forest.[12] It uses distributed ledger technology to allow nonhuman entities—in this case, a forest in Germany—to effectively own themselves. This is possible because ledger technology enables the sharing, synchronizing, and replicating of digital data across multiple and disparate geographical sites, without a central administrator or centralized data storage. terra0 uses a specific kind of blockchain, called Ethereum, to allow the forest to become an autonomous, self-utilizing land parcel.

The forest manages itself according to rules embedded in its blockchain-enabled "forest surveillance smart contract," which is hosted on a server named the Oracle. The contract specifies the times at which logging licenses are sold, at a price determined by an automated algorithm that monitors the timber market and learns from past sales to set timber prices. As logs are sold, the forest automatically accumulates capital in the form of wood tokens (a blockchain-based cryptocurrency). Eventually, self-utilization enables the forest to buy itself back from human landowners, and perhaps even expand its land base. The forest, as a technologically augmented ecosystem, is thus also an economic actor. The forest could own a bank account, collect revenues, and sell carbon offsets (while accounting for the energy generated by its digital systems). The digitally enhanced forest ecosystem is now a nonhuman economic actor, able to enter into transactions just like humans. Could environmental organizations like the Nature Conservancy loan money to the forest to purchase itself, which it would eventually pay back through proceeds generated from sustainable forestry?

One can imagine combining such a system with autonomous environmental sensor networks and automated self-monitoring algorithms that would allow the forest to assess its own ecological health in order to guide its timber sales, enabling the forest to shape its own destiny, presumably in line with assumptions supporting regenerative sustainability encoded by the algorithms' creators. Sensors would be embedded in tree trunks and the soil, and on the animals, birds, and insects that move through the forest; drones and satellites would monitor from overhead; machine learning algorithms would analyze the data to produce real-time reports on the forest's health, and even the mood of its denizens. Ecosystem services could be quantified and measured; the various forest creatures (from fungi to insects, rodents to trees) could be recognized as ecosystem workers, providing valuable services to one another, not just humans. If proposals to use blockchain to create interspecies money come to fruition, a mechanism would exist via which digital payments could be disbursed to terra0 (or, more precisely, its digital twin).[13]

terra0's tagline ("On the blockchain, no one knows you're a forest") proposes a quasi-libertarian vision of nonhuman agency. Does automating ecological resilience help to combat speciesism? My intention in sharing this example is not to advocate for this approach. There are many complicated issues that would need to be solved, such as jurisdiction and the balancing of competing claims in the ecosystem. For example, if terra0 were to be brought to life, how would it adjudicate between what is good for the trees versus what is good for the birds or the mycorrhizal fungi in the forest soils? Where would its borders lie, at the edge of the tree canopy or at the edge of the (much wider) root network? Would longer-lived species (like trees) have more influence than shorter-lived species (like honeybees)? Moreover, property rights for nonhumans may be at odds with Indigenous property rights, or systems of common property rights. More fundamentally, property rights are a culturally specific system of land ownership that is aligned with capitalism and colonialism, and creates a zero-sum game: own or be owned, dominate or be dominated. A more equitable approach might require new forms of relational property rights—such as easements—that enable many creatures and species to share space, with mutual rights and responsibilities, and without any one entity being an owner.

And even if blockchain provides a viable mechanism for nonhumans of various kinds to own themselves, another question arises: *Should* we use this approach? This is a tricky question to answer. Blockchain is not a neutral technology from a political economic perspective, nor are property rights to digital data.[14] Indeed, property rights have been a major mechanism of "accumulation by dispossession," colonialism, and environmentally destructive resource extraction.[15] As Cambridge University sociologist Jennifer Gabrys argues in her cautionary analysis of the "Internet of Trees," we have not yet fully thought through the political implications of our technological choices.[16]

Is extending private property ownership to nonhumans through blockchain a viable way to advance conservation goals? One can imagine the criticisms directed at cryptocurrency also being leveled at GainForest and terra0.[17] Will the development of blockchain applications for conservation follow a more capitalist model (like Bitcoin), a not-for-profit model (like Farm Share, FairCoop, and Bitnation), or a cooperative model (Nature 2.0, funded by the creator of Ethereum, an initiative that seeks to disrupt environmental governance with blockchain)? These and other blockchain-based options are currently proliferating, and it is unclear which one will win out. Embedding responsibilities that reflect reciprocal relationships may alleviate some of these concerns, but it is not yet clear whether blockchain offers a progressive or a regressive pathway for nonhuman agency. Moreover, concerns have been raised regarding the community-related impacts of Bitcoin mining; crypto-mining operations often target communities with cheap electricity and permissive regulations, driving up energy costs, placing strain on residential power grids not designed for such high demands, and even creating fire hazards through spikes in electricity use that surpass safety thresholds in residential wiring.[18] These concerns have led some cities in the United States to ban crypto mining. The climate emissions created by blockchain are also enormous, with Bitcoin alone using as much electricity as the entire country of Sweden, although (at the time of writing) the decision to move Ethereum from "proof of work" to "proof of stake" protocol has reportedly reduced its carbon emissions by 99 percent. Nonetheless, from an environmental perspective, carbon emissions might be the Achilles' heel of digital ledger technology innovations.[19]

I have discussed property rights for nonhuman living beings, and discussed how digital technology could aid us in this endeavor. But what about digital technology itself? On which grounds could a robot or an algorithm be recognized as a legal person? A variety of criteria have been put forward for recognizing animals as legal persons: some capacity for communication, a sense of self, purposive intention to achieve one's plan in life, and the ability to live in a community with other organisms based on mutual self-interest, to name just a few.[20] Advocates further assert that animals possess inviolable rights, as animals have their own subjective existence and moral significance, which must be respected. The conception of moral equality underpinning the doctrine of human rights should, from this perspective, simply be extended to animals.[21]

Although similar arguments have been made regarding the rights of algorithms and robots, few involved in digital rights debates have engaged with animal rights, and vice versa.[22] And the debates appear to differ in their emphasis: the emphasis on intelligence as a precondition for rights is, for example, stronger in the AI rights community than in the animal rights community.[23] But these debates raise a common set of questions: Should nonhumans have legal standing? Are they rights-holding subjects? Should they be granted legal personhood? If you grant rights to a river, why not to a robot? In my view, rivers and robots are not equivalent. The former exists in reciprocal relationship with Gaia; rivers provide ecosystem services, such as habitat, climate regulation through the hydrological cycle, and the dilution of pollution. The latter exists in a parasitic relationship, as digital technology requires mining resources, with attendant environmental and human impacts, carbon emissions, and e-waste.

The parallel questions that I have raised here about the rights of animals and algorithms raise issues that, in the future, will be increasingly entangled. Proponents of blockchain conservation and nonhuman rights for animals should ask whether the innovations they advance may also set precedents for rights to be extended to machine entities such as robots and algorithms. These two debates are occurring in parallel, with little overlap. Yet they have many similarities, and there is a concept that may bridge the two debates: symbiotic autonomy.

SYMBIOTIC AUTONOMY: COLLABORATING WITH ALGORITHMS AND ANIMALS

Manuela Veloso was frustrated. Her robot was stuck. It had a simple task: carry an object down a hallway from one part of the computer science building to another. But strong sunlight and shadows brought the robot to a confused standstill. Despite being outfitted with the most expensive hardware she could afford, it couldn't even open doorknobs or avoid simple obstacles. The last straw: her team told her they'd have to buy another arm to help the robot take the elevator, as its existing arms couldn't push buttons in the right way. Veloso, a professor of computer science at Carnegie Mellon, had been named one of the most influential AI scientists on the planet. But even after several decades of research, she could only produce robots that had to be constantly trailed by graduate students, like clumsy, overtired toddlers followed by a group of nervous babysitters.

Veloso's frustration reflects a conundrum in contemporary robotics. Critics like MIT professor David Mindell note that autonomous robots are safe but underwhelming in terms of performance: "They don't do much collaboration, but at least they won't cut your head off."[24] Gary Marcus, a neuroscientist who has spent his career at the forefront of AI research, pokes fun at robots: "Worried about super-intelligent robots rising up and attacking us?" writes Marcus. "Close your doors, and for good measure, lock them. Contemporary robots struggle greatly with doorknobs, sometimes even falling over as they try to open them. Still worried? Paint your doorknob black, against a black background, which will greatly reduce the chance that the robot will even be able to see it."[25] As Marcus's tongue-in-cheek scenario notes, the world's most sophisticated robots are defeated by doorknobs, as well as banana peels, staircases, and shadows.

One night, Veloso woke up from a dream with an epiphany. Roboticists might never reach the goal of a completely humanlike robot. The goal that had lured her into computer science—creating robots exactly like humans—was unachievable. But this goal, to which she had devoted her entire professional career, actually didn't matter. "Forget it," she remembers telling herself. "We have to embrace the fact that robots will have limitations. Robots will always have constraints compared with living organisms." But she had a radical idea about how to solve this problem. "What we need," she told herself, "is for robots to recognize their limitations and autonomously ask for help."[26]

The idea seemed simple. Robots have limitations, but they also have lots of resources: the internet, other robots, and humans. If robots could draw on these resources, they could achieve their goals through collaboration. Rather than buying an expensive arm for pushing elevator buttons, she reasoned, she could simply code the robot to ask a nearby human for help. Once robots could ask humans for help, they—and the entire field of robotics—would take a giant leap forward.

Technically, her plan was flawless. But Veloso forgot one thing: other humans. When she shared her idea with her graduate students, they were resistant. "This seems like cheating, or giving up," was a common pushback. Curious, Veloso dug deeper. Why was it so hard to accept that a robot could ask for help? She had run up against a strongly held belief in computer science: building human knowledge into robots or artificial intelligence systems is viewed as cheating (at worst) or simply undesirable (at best).[27] The mainstream, conventional view in AI holds that algorithms should learn on their own; in other words, AI should avoid external inputs, like assistance from humans, and only computation, not human knowledge, should be leveraged by robots. From this perspective, to be autonomous means to be alone: self-reliant and solitary. The label given to this idea, "strong AI," refers to an AI that learns from scratch, eventually teaching itself to solve new problems, and for many computer scientists, this form of autonomy is the goal.

Another reason for the resistance of Veloso's fellow computer scientists was psychological. As she observes, "In the past, we built robots to service humans. We presumed their sole purpose was to provide services to meet human needs, by executing commands. Because of this, the idea that robots could ask for help from humans didn't exist."[28] Robots were servants, not colleagues, coworkers, or friends. This objection was even harder to overcome, yet despite initial resistance, Veloso eventually won over the skeptics. Her first radical step: coding humility into the robots. Her team began by testing a new generation of collaborative robots (christened Cobots). They first taught the Cobots—which look like iPads on wheeled pedestals—to say "I don't know" in response to commands or situations they didn't recognize. As Veloso argues: "We should embrace the idea that digital artifacts should recognize when a situation does not meet their expectations; they should exhibit self-awareness and ask us to explain our world to them." This reduces the chance for mistakes or algorithmic bias, due to the emphasis on solving problems even with low confidence. Rather than saying, "I think this is an apple with 20 percent confidence," a computer could simply say, "I don't know what this is, please tell me." Programming the robots to be humble presented an interesting technical challenge: the robot had (1) to recognize the need for actions it could not itself execute and (2) to generate sequences of actions that are impossible without external assistance. The environment becomes a new source of inputs for collaborative problem-solving with the robot, in an iterative loop.

Veloso didn't stop there. Her next step: programming vulnerability into the robots, by teaching the Cobots to ask for help. This makes sense, Veloso argues, because collaborative robots have physical bodies that encounter all sorts of obstacles. Self-contained "strong AI" makes sense for completely computational tasks—like recognizing photos or playing chess or Go—but when a robot picks up your child or takes an elevator, it may sometimes need help—just as some humans do. To the team's surprise, even human strangers responded well to the Cobots' requests for assistance. Humans would press elevator buttons, hold open doors, or put things in the Cobots' baskets. Soon, the Cobots were roaming the building on their own. Students no longer had to follow them around. The Cobots even began showing up at Veloso's office carrying new items; no one knew how they had gotten there. They even ran errands for other people. When Veloso abandoned the notion that robots are solitary, she ironically achieved their breakthrough goal: helping the robots to become more autonomous. However, this was not achieved through computational supremacy. Rather, Veloso's collaborative robots are more adaptive and autonomous because they have been designed to be humble, reflexive, vulnerable, and interactive, all of which help them to be better than the average robot at relationship-building.

If AI is to prove useful in mobile robots that share our living and working spaces, Veloso is convinced that her humble, vulnerable robots will lead the way. "It's an inevitable fact that we are going to have machines, artificial creatures, that will be a part of our daily life," says Veloso. We are used to having machine intelligence in our smartphones; soon, digital intelligence will have a body, and move around with us. And because intelligence is embodied, it needs to be both collaborative and social. "When you start accepting robots around you, like a third species, along with pets and humans, you want to relate to them." Veloso believes that we need to begin building relationships between humans, nonhuman animals, and social robots, which will one day be ubiquitous. Her name for these relationships is "symbiotic autonomy," the mutual coexistence of living beings, robots, and software in an ever-deepening interdependence. Symbiotic autonomy, she argues, is an emergent property of our innovation frontier. In the future, humans and AI-powered robots will cohabit, linked via a continual exchange of information.

Luciano Floridi, a professor at the University of Oxford's Internet Institute, argues that Veloso's insight rests on four profound changes in the human relationship to the world in a hyperconnected era: the blurring of the divide between reality and virtuality; the porosity of boundaries separating human, machine, and nature; a reversal of information availability, from scarcity to abundance; and a shift from the Internet of Things to the Internet of Experiences.

If correct, what does this imply for environmental governance in the future? Concepts of interdependence, community, and symbiosis are conventionally reserved to speak about relationships between living things. But what if our symbiotic future includes not only living organisms but also robots and intelligent machine systems? As explored in the following chapters, Veloso's concept of symbiotic autonomy anticipates hybridization and fusion, an emergent set of relationships between beings as diverse as robots and plants, insects and clouds.

Within ecology, scientists define symbiosis as a long-term, intimate association of two dissimilar organisms, often of different species. In the past, symbiosis was presumed to be purely biological. Veloso's claim seems to imply that symbiosis can also be technological, because technology and ecology are increasingly incorporated into symbiotic relationships across extended networks of organisms and species. Philosopher of science Donna Haraway offers another perspective on these issues through her concept of sympoeisis ("making-with").[29] In Haraway's view, we should embrace the idea that different beings—animal and machine—are intermingled biophysically as well as technologically. Haraway calls these techo-natural creations "cyborgs." The biologist Lynn Margulis used the term "holobiont," meaning symbiotic assemblages that create thriving beings—which are simultaneously communities. Both these terms seek to undermine the individual-as-organism view on which Western science is predicated and show that we need a new way of thinking about interrelationships and interspecies being in an era of ecological crisis.

Sympoiesis draws our attention to the collective, self-organizing nature of living systems, which are composed of multiple entities that interact, producing emergent patterns and structures. Bacteria, plants, humans, and machines all interact, for example, to produce the global atmosphere. When forests breathe out, we breathe in. Living beings are not discrete, bounded entities; metabolically, we exchange nutrients and energy with other beings on Earth. We are thus constantly entangled with other living beings and machines in a web of interdependence. Jason Lewis, an Indigenous scholar of computation arts, calls this "making kin with the machine."

These concepts all invoke a sense of reciprocity, collaboration, and perhaps even reconciliation. They share a common theme: acknowledging nonhumans as subjects that might be legal persons, might have property rights, and might collaborate with humans in governing the planet as a shared home. Digital technologies now play a crucial role in mediating these relationships. Humanity, including its technologies, is part of nature and exists within a web of relationships between living beings and machines. But this is not necessarily emancipatory, as these webs of interdependence often contain deep power imbalances, and they are not always mutually beneficial. Humanity already lives in a one-sided relationship with many other organisms: our relationship with other species is often characterized by predation, competition, and parasitism. A spider and a fly both encounter the spider's web, but the outcome is very different for the latter.

There are, of course, symbiotic relationships that do not involve harm: commensalism (one organism benefits from the symbiotic association, without harming the other) and mutualism (both organisms benefit). A critical question for Digital Earth governance is whether and how (bio)digital technologies could be deployed to support and enhance mutualism—enhancing the habitability and sustainability of life on Earth for humans and other species. Yet in the absence of equal rights, mutualism is likely to give way to predation or domination. It seems desirable, and perhaps necessary, to consider whether nonhumans could hold legal rights, under some conditions. Symbiotic autonomy in the absence of such rights runs the risk of masking exploitation couched in deceptive terms: it implies a degree of agency for the subaltern, but may be no more than a projection of a fantasy by those who hold power. French philosopher Michel Foucault coined a term for this fantasy: biopolitics, or the administration of life through a society of surveillance so ubiquitous that obedience is internalized. If digital technologies become biopolitical, the autonomy of nonhumans may be suppressed rather than enhanced, and they may be further subjugated to human desires.

What might counter this biopolitical impulse? The following chapter turns to this question and explores whether digital technology could help us cultivate greater empathy for other species. Could digital technologies, so often associated with alienation from nature, enable humanity to reconnect instead?

PARABLE OF BIIDAABAN

In the augmented reality, like a waking dream, the viewer walks the streets of a future city. Vines crumble sidewalks. The streetlights are extinguished, and the sky is dense with stars. Mold coats the walls as she descends into a derelict subway station. The Indigenous languages of the area (Anishinaabemowin, Kanyen'kéha, Wyandot) echo along the tunnel walls. The words have grown from the land in the same way that the plants do. At the end of the tunnel, the darkness lifts, and final words are spoken: *Biidaaban*, *Anishinaabemowin* for the moment of the first light of dawn, when past and future collapse into eternal present.

When she takes off the goggles, the viewer carries the vision with her. The app tracking her behavior registers a change in subsequent days. She brings some water to the thirsty trees struggling on the sidewalks. Stops to watch the hawks hunting from the tallest rooftops. Sees the footprints of a coyote at the edge of the ravine under the bridge as she walks to work.

She had never really noticed them. She had forgotten that she had forgotten. She is remembering how to remember the future.

8 EMPATHY MACHINES

You are standing in a dim forest. Infrared markers are strapped to your arms and affixed to your back. Donning a helmet, you select a dragonfly avatar. Suddenly, the sounds of the forest come alive, and the colors intensify: the forest is no longer drab, but alive with light. You are now "seeing" like a dragonfly. Your eyes see flowers glimmering in the ultraviolet spectrum as well as familiar shades of red, green, and blue. Because dragonfly eyes take five times as many frames per second as humans, time appears to slow down. Mosquitoes drift by like slow-moving clouds. You can feel forest sounds through your body, just as the dragonfly does. And you can fly. Your wearable subwoofer backpack thrums on your shoulder blades, mimicking the sensation of beating wings.

Welcome to *In the Eyes of the Animal.* Set in England's Grizedale Forest, and developed by the artist collective Marshmallow Laser Feast, the augmented reality (AR) experience aims to enable humans to experience the environment as forest animals do.

In the augmented forest, you can fly like an insect or bird; advances in haptic (touch) technology mean that you really feel as if you are flapping wings and flying. Why is this significant? AR innovators argue that hacking our physical senses (particularly our sense of touch) creates a sense of inhabiting nonhuman bodies. Their ultimate goal is to enable humans to completely immerse themselves in another animal's perceptual way of being in the world (*umwelt*), by stimulating all of our senses: sight, touch, sound, smell, and even taste.

Why would immersing oneself in the world of a nonhuman being be useful in countering environmental degradation? Proponents of immersive technologies (which include augmented, virtual, and mixed reality) argue that they help us to experience the world in new ways, and often from the perspective of others, thereby cultivating empathy with the nonhuman world. Traditional tools for cultivating empathy encourage us to *see* ourselves in others. Immersive technologies go one step further: they attempt to cultivate a sense of *being* oneself in another. Advocates hope that this might spark a deeper, more authentic desire to save the planet.

Why would these immersive technologies have such powerful effects? Advocates argue that VR achieves a sense of presence. As VR artist Chris Milk puts it, VR helps us walk "through the window," rather than merely watching the window. Some experimental evidence indicates that the brain stores and accesses memories of VR just as it does real-world experiences; in other words, your brain believes that the virtual environment is real and that you are actually existing within it. As *Wired* editor Peter Rubin argues, VR thus elicits a much broader range of embodied physical and emotional reactions than any other digital medium. The result is a profound and unprecedented sense of intimacy. In other words, proponents argue that VR and AR give us a sense of "withness" that is distinct from merely witnessing events via video or reading about them.

One of the leading proponents of immersive technology as a mechanism for increasing human empathy for the environment is Jeremy Bailenson, who calls himself a "virtual reality designer for social good." A professor in Stanford University's communications department with a PhD in cognitive psychology, and founding director of the Virtual Human Interaction Lab, Bailenson studies how immersive virtual experiences can affect psychological processes, including empathy, bias, and decision-making. In his immersive VR experience *Becoming Homeless*, viewers spend days in the life of someone who can no longer afford a home, trying to protect their remaining belongings while living on the street. Viewers who participated in Bailenson's immersive VR design had more positive attitudes about the homeless and were significantly more likely to sign a petition in support of affordable housing for the homeless.[1]

One of Bailenson's first environmental VR experiences was inspired by a *New York Times* article that explained how the rapidly growing use of soft, fluffy toilet paper in the United States was leading to the unnecessary death of millions of trees, including old-growth forests. Although many Americans say they are concerned about the environment, less than 3 percent of them use recycled toilet paper. Curious, Bailenson put his team to work. One of his graduate students built a beautiful virtual forest, with towering sequoia trees and singing birds. Then she designed the virtual reality experience. After viewers donned VR goggles in Bailenson's lab, a virtual chainsaw was placed in their hands; within the simulation, they cut down a giant tree. After it smashed to the ground, viewers were asked to walk around and examine the damage in an eerily quiet forest: the birds had been silenced. After viewers removed the VR headsets, they were told that using nonrecycled toilet paper would lead to the death of at least two such giant trees in their lifetime. At that point, a graduate student was primed to walk by, knock over a glass of water as if by accident, and ask the viewers to help her clean up the spill. Those who had participated in the VR experience used 20 percent fewer napkins than a control group that had watched a video or read a description of the trees being cut.[2] The effects persisted after participants left the lab.

In one experiment after another, Bailenson's results are consistent: virtual reality experiences that cultivate a sense of "self as other" result in changes in environmental attitudes and environmental behaviors. Behavioral change is most significant when participants in a VR experience actually move through a physical landscape, which heightens the "whole body" sense of immersion. Video watchers don't change their behavior as much. Simply reading text appears to have little or no effect. Within a few hours of reading the opening paragraphs of this chapter, you will likely have forgotten my story, but if you participate in one of Bailenson's virtual reality experiences, not only will you remember, but you will be more likely to act.

Bailenson has now embarked on what he terms the world's largest virtual reality experiment. The Stanford Ocean Acidification Experience is an underwater virtual reality experience that allows you to observe the future decline of coral reefs over the coming century, translating marine science research into a popular VR film that premiered at the Tribeca Film Festival.

Within the VR simulation, viewers watch as the ocean absorbs normally invisible molecules of carbon dioxide; time speeds up and the coral reef degrades in mere minutes, as its once abundant marine life dwindles and disappears. In one simulation, high school seniors took on new virtual identities: bright pink coral on a reef pulsing with fish, mollusks, and other marine life. As the simulation unfolds, the fish die off and are replaced by slimy green algae. By the end of the simulation, the viewer's virtual coral skeleton simply disintegrates as the narrator declares: "If ocean acidification continues, ecosystems like your rocky reef, a world that was once full of biological diversity, will become a world of weeds."[3] The Ocean Acidification VR experience produced some of the most powerful changes in viewers that Bailenson had observed.[4] So he decided to scale up: creating sixteen different versions, and presenting the VR simulation to decision-makers around the world, including the US Senate. It has now been translated into multiple languages and downloaded in over 100 countries. Says Bailenson: "Changing the right minds can have a huge impact."[5]

Bailenson's work is an example of a new generation of digital technologies that are being mobilized to enhance awareness of, and empathy for, nonhuman species. Proponents argue that new forms of Digital Earth governance will be viable only if our perceptions of, and attitudes toward, other species undergo a shift: more biocentric, less human-centered. Scientists, educators, and activists have developed VR, mixed reality (MR), and AR applications in an attempt to immerse humans in the *umwelt* of other beings. There are two assumptions behind this work: these technologies will cultivate some kind of understanding or fellow feeling on the part of humans for other species, and for nature more broadly, and this feeling will lead to more environmentally sustainable behaviors.

But do virtual and augmented reality experiences really change human attitudes and environmental behaviors? The results are mixed. One important variable seems to be the incorporation of realistic physical movement, although researchers aren't sure why this is the case. Cyberculture innovators have tended to overlook biological reality when exploring virtual reality, but it appears that blending the two is more successful in changing people's opinions and actions.[6]

And while Bailenson's Virtual Human Interaction Lab has consistently shown changed behavior after VR experiences, empathy is harder to measure (and scientists dispute how to define it). Participants in VR experiences are more likely to take the perspective of someone else (one facet of empathy), but only the specific person whose perspective they assumed in the simulation.[7] Experiencing a VR simulation, in other words, does not seem to enhance generalized empathy. This finding is in line with Bailenson's other work, on the "Proteus effect" (after the shape-changing Greek god Proteus): digital avatars change people's behavior offline as well as online.[8] Our mutable selves change their opinions and behaviors in the digital world and the real world.

This discussion of empathy also highlights a blind spot: a focus on the emotional experience of the human viewer, rather than the object of their gaze. Mounting scientific evidence confirms that a wide variety of species exhibit emotional states.[9] If this is true, then VR and AR experiences are one-sided, insofar as they seek to stimulate an emotional response in humans without incorporating insights about the emotions of the nonhumans they seek to simulate. This may change in the future. By using computer vision and machine learning, scientists can decode subtle emotions on the part of animals. In one experiment, researchers used an unsupervised machine learning algorithm to classify a wide range of mouse emotions, such as disgust, active fear, passive fear, pleasure, or curiosity.[10] Could VR and AR incorporate these types of insights? One can imagine the equivalent of a Google Translate for emotions, enabling real-time sensing of another creature's emotional states. But one can equally imagine a VR or AR system that interprets these emotions incorrectly, creating a self-referential and insidiously fictitious experience.

VR and AR experiences are also culturally myopic, likely to reflect the biases of their designers, who are likely to be WEIRD (from Western, educated, industrialized, rich, and democratic countries).[11] Environmental categorization and reasoning vary significantly across cultures. Some of Bailenson's work backs this up: in one study, he demonstrated how Itzá Mayan villagers in Guatemala offered nuanced categorizations of birds, employing causal and ecological reasoning rather than the taxonomic reasoning used by US undergraduate students.[12] VR design is culturally mediated, and hence potentially culturally biased. Without inclusive design, exclusionary practices that have long plagued the environmental movement, as well as the earth and environmental sciences, may be replicated.

VR GOES GREEN

Bailenson is not alone in seeking to use VR to advance the environmental movement.[13] Animal rights activists have created visceral VR experiences of slaughterhouses and factory farms. Environmental organizations (and even mainstream media organizations like *Time*) have created VR apps to help raise awareness of the impacts of environmental degradation and climate change on the Amazon rainforest. There are VR apps for our most iconic ecosystems, from the Amazon to the Arctic, and also for the hardest-to-reach places on the planet, such as the deepest oceans and subterranean Earth. Building on the success of VR documentaries like *This Is Climate Change* and *Greenland Melting*, the Sierra Club has christened VR the "secret sauce" of climate change action. VR experiences are now widely available as virtual field trips, taking viewers to destinations like Yosemite, Antarctica, and Mount Everest.

Augmented reality apps like After Ice go one step further. Hold your phone up in front of you, and the app shows how much of the landscape will be underwater by 2080—a simple yet powerful way to connect the individual experience to abstract scientific predictions of sea level rise. Scientists have even developed mobile AR devices for environmental monitoring and management, like flood visualization apps you can download to your phone, which show you how flooding would affect the banks of a river on which you are standing.

Museums, zoos, and schools are also experimenting with augmented reality. To raise awareness about climate change, Coca-Cola funded an augmented reality video of a polar bear and its cubs, separated by fast-melting Arctic ice floes, in the heart of the Science Museum in London. To raise awareness about the endangered Caucasian leopard, the World Wildlife Fund collaborated with an Armenian company to develop an AR leopard that roamed the Yerevan Botanical Garden, sitting, pacing, and even roaring—while letting people pose with him, pet him, or even take a selfie with him. You can download a free, pocket-sized version and snap selfies with the leopard wherever you go. Some critics argue that such AR experiences may encourage users to be overly comfortable with fictitious nature experiences, and will thus lead them to act inappropriately in real-life situations. But such criticisms do not appear to have slowed the wave of AR experiences now being developed for environmental purposes.

Belying the early predictions of naysayers in response to glitchy 1990s technology, VR and AR are now going mainstream. Hundreds of virtual, augmented, and mixed reality experiences are produced every year, at hackathons and extended reality expos like EarthXR and the World XR Forum. The deployment of 5G wireless networks promises to enable streaming of VR and AR from the cloud, lowering the cost of headsets and making the experiences even more realistic.

The next step in the extended reality journey, according to advocates, is bridging the human/nonhuman divide. Could augmented reality provide a convincing experience of actually *being* another species? Many environmental organizations are now trying to achieve just that. The Rainforest Alliance's *Tree*, for example, invites viewers to transform themselves into a rainforest tree: "with your arms as branches and your body as the trunk, you'll experience the tree's growth from a seedling into its fullest form and witness its fate firsthand." Some of the most intriguing VR experiments have come about through collaborations with artists.[14] In the Saatchi Gallery's "We Live in an Ocean of Air" exhibit, for example, spectators roam around the prestigious gallery in the heart of London wearing augmented reality headsets. The AR experience transports them to an ancient forest of giant sequoia trees, which promises to reveal the invisible connection between plant and human—through breath. Like many environmental AR experiences, the designers strive to foster a mental merging of the self and nature.[15]

THE MAGIC WELL

Will VR and AR experiences one day be ubiquitous? A few years ago, tech sector hype reached a fever pitch. A new term entered the lexicon: mixed reality, a merging of virtual and biophysical worlds in which physical and digital objects coexist and interact in real time. Tech enthusiasts also referred to this phenomenon as the Mirrorworld, a future in which augmented reality layers information over the real world. Built using spatial computing, which enables digital objects to be situated in physical spaces, the Mirrorworld houses digital twins—online replicas of physical entities—side by side with their real-world counterparts. If this is hard to imagine, it's because it doesn't exist yet. Pokémon GO was a foretaste. But instead of a few Pokémon scattered here and there, imagine that everything in your environment has been digitally tagged, enhanced, and annotated. If the enthusiasts are to be believed, mixed reality will be universal.

The rise of Magic Leap is an illustrative example that casts some doubt on this claim. Magic Leap was a tech start-up working on the next generation of augmented reality, backed by Google, Apple, and Microsoft and supported by several billion dollars of venture capital funding. The company's founder, Rony Abovitz (Twitter bio: "friend of people, animals, and robots"), promised incredible hardware that would turn the entire planet into an artistic medium, a digital blank canvas waiting to be filled. *Wired* magazine compared Magic Leap's initial product to the Matrix or *Snow Crash*'s metaverse. Abovitz even won over professional skeptics like Adam Savage, host of the TV show *MythBusters*, who described his emotional response to a whale swimming by his office windows in the metaverse: "I actually got choked up."[16]

The Nobel Prize–winning scientist Karl von Frisch had a name for Savage's feeling: the "magic well," a profound feeling of transcendence when humans are present with nonhumans.[17] As humanity has colonized the planet and eradicated wild animals across much of the planet, the magic well has been harder to find. But techno-prophets like Rony Abovitz argue that planetary computerization will make this feeling pervasive, even universally accessible. Imagine Savage's virtual whale swimming by your bedroom window. In other Magic Leap experiences, you can hike or scuba-dive in your house (with an optional entourage of virtual nonhumans or even supernatural beings). As the Mirrorworld comes into being, Abovitz argued, the transcendent feeling of being fully present with nonhumans will be easily accessible to all of us.

Humans will not necessarily need to venture into the environment like the explorers of yesteryear. Rather, we will be able to immerse ourselves in the environment anywhere, anytime: augmented reality will bring nature to you, in the comfort of your couch. Abovitz called this the Magicverse: a natural successor to the World Wide Web. Instead of accessing the internet through a screen, any and every surface on the planet becomes a window into the spatial digital world, with mixed reality experiences continuously available and highly personalizable. Critics argued that Magic Leap was overhyped, and predicted that it would fail just like Google Glass, pointing out that much of the necessary technology has yet to be built (not least the expansion of 5G and new 6G wireless technology required to support the augmented reality cloud). Magic Leap did stumble in 2020, laying off 1,000 workers as its CEO stepped down.[18] The company seemed to have lost its big bet on consumer mixed reality; at the time of writing, the company had released its latest AR headset while some of the business press was predicting that it would be become a niche business offering a product akin to Microsoft's HoloLens, it was also in discussion with Meta, Facebook's parent company to counter Apple's Vision Pro. Still, at least for the foreseeable future, VR and AR experiences appear unlikely to become widespread.

"FAKE" VERSUS "REAL" NATURE

In recent years, environmentalists have embraced social media.[19] Environmental groups solicit "likes" from their supporters and urge the public to share environmental news, campaign slogans, and requests for donations. You can now go camping on Twitter, mountaineering on Snapchat, trekking in Google Street View, or take action online with groups like the Nature Conservancy, World Wildlife Fund, and Conservation International. A new generation of "independent activists" tell their first-person stories without relying on traditional intermediaries like large media or environmental organizations. Geolocation enables anyone, anywhere, anytime to highlight environmental injustice—such as when #droughtshaming went viral during the 2015 California drought. Social media campaigns are also a well-honed tactic used by environmental groups, sometimes to great effect. For example, in its campaign to curb Shell Oil's activities in the Arctic Circle, Greenpeace targeted Shell's marketing partner LEGO with a video depicting a pristine Arctic (built with over 200 pounds of LEGO) being covered in an oil spill. The video went viral (over 9 million views) and LEGO reluctantly canceled its partnership with Shell. And a variety of animals now post on Facebook and Twitter (@MaryLeeShark and @EstherThePig have over 100,000 followers each, and Tatum the Talking Dog has over 3.5 million), and conservation groups are experimenting with data-to-text natural language generation (NLG) algorithms to generate blog posts communicating ecological insights, based on satellite tag data from birds.

But can nature be saved through mere mouse clicks, shares, and retweets of viral videos? Proponents argue that so-called clicktivism is highly effective at disseminating information and mobilizing the "crowd" to pressure corporations and governments to take action. But critics argue that clicktivism is a couch-potato form of pseudo-protest that distracts from, or perhaps even inhibits, meaningful action. Individual clicks may be misleading, as they aren't necessarily correlated with long-term commitment. Even worse, they may feed us fake environmental news, or even assuage our environmental guilt or lull us into inaction by providing a pretense of engagement. As environmentalism goes digital, humans might be losing our connection with the environment—while suffering from the illusion that we are still connected.[20] Clicktivism might be better termed "slacktivism."

If social media further separates us from the natural world, could augmented and extended reality have the opposite effect? Could the shift toward virtual reality and experiences of digital nonhumans enable us to reconnect to the natural world in a more authentic way? Could humanity derive some benefit from immersive connections in nature without actually being there, or is this a dangerous illusion? These questions are often posed by environmentalists who distrust virtual and augmented reality. Eco-philosophers like Richard Louv (author of the bestselling *Last Child in the Woods* and *The Nature Principle*) argue that the only authentic nature immersive experiences in nature are nondigital. Nothing can replace direct experience in real nature, with as little technological mediation as possible. From this perspective, physically connecting with real nature "out there" has no substitute.[21]

Research shows that they are correct: virtual nature experiences are not quite the same as the real thing. Nevertheless, even virtual nature experiences may have a beneficial effect. Peter Kahn, an eco-psychologist, has crafted ingenious experiments that measure the psychological and physiological aspects of our technologically mediated interactions with nature.[22] In one experiment, Kahn placed "technological nature windows" (large plasma screens broadcasting real-time views of nature) in offices on campus, then analyzed the effects on viewers. He also studied online nature activities like "telegardening" (which enables online users remotely controlling a robotic arm to plant seeds and tend a real garden), and the relationships of both children and adults with robotic pets (like the AIBO robotic dogs). Other researchers have studied whether nature videos played in prisons reduce prisoner violence (they do), whether watching *Planet Earth* reduces anxiety (it does), and whether workers in offices with plasma screen "windows" that play livestreams of the actual outdoors are happier and more productive (they are).

Kahn's work has a good news/bad news implication for advocates of virtual and augmented reality. Experiencing technologically mediated nature is better than no nature. But it's still not as good as real nature—at least not yet. Kahn argues that the human desire to connect with nature is hardwired, that there is a deep-rooted need for nature. But, Kahn argues, virtual substitutes don't have the same effects. As he puts it, "we should develop and use technological nature as a bonus on life, not as its substitute."[23]

THE SPELL OF REAL NATURE

Although he didn't test the latest generation of VR/AR technologies, Kahn's findings are in line with the views of many influential environmentalists. As the much-cited environmental thinker David Abram writes in the postscript to his classic environmental text *The Spell of the Sensuous*, humans thrive in landscapes that make us feel alive. And we feel most alive and fully human when we interact with nonhumans. But, as Abram warns, "with these new technologies, there's no real otherness there. As we enter into a field of ubiquitous computing and the Internet of 'Things that Think,' we mistakenly seal ourselves into an exclusively human field of interactions."[24]

Rather than experiencing nonhuman worlds, we are actually just talking among ourselves. Rather than an *inter*-species internet, we are creating an *intra*-species internet: VR experiences are mere projections of human desires, fantasies, and fears. In other words, the "nature" we experience through VR and AR is fictitious—and hence second-rate. To the extent that it is delusional, because we mistake it for the real thing, it may even be dangerous.

According to Abram, humanity is most fully expressed when we relate deeply and directly to the nonhuman—outdoors, through all our senses, and unmediated by technology. Virtual and augmented reality do the opposite: they have us relate, instead, to a human projection of the nonhuman. Essentially, AR and VR experiences are insidious, narcissistic fantasies. Think about the vignette with which this chapter opened: wandering through a forest with a headset that renders you blind and deaf to the world around you, focused on computerized simulacra. You have the impression that you are experiencing the forest; meanwhile, a real fox might trot by, and a real dragonfly might hover right beside you—experiences to which you have blinded yourself. VR can't substitute for locking eyes with a real deer that you surprise on a forest path, for watching a bear in a stream catching spawning salmon, or watching a spider spinning its web and wrapping a cocoon of silk around its prey. Once you've had these sorts of experiences—and they are almost impossible to describe to those who haven't without sounding maudlin, mystical, or slightly unhinged—you know that critics of virtual and augmented reality are right. People who have transcendent experiences of nature inhabit, at least for a short while, an enchanted space. Compared with these intimate moments of nature's presence, some environmentalists argue, the digital thrall is both a distraction and a menace.

Another critique can be leveled at virtual and augmented reality from a biophysical perspective. Imagine walking in a pine forest. Think of the smell the trees make: the sharp, outdoorsy, unmistakable Christmas tree smell. What makes that smell? Terpenes: chemical compounds that act like signals. When insects attack trees, the trees release volatile organic compounds, like terpenes, into the air. Wafting through the air, these herbivore-induced plant volatiles (HIPVs) attract the attention of a variety of insect predators and enemies: spiders and birds that eat insects, or parasitic nematodes that carry insect-killing bacteria in their bodies. The trees use terpenes and other biochemical signals to communicate.[25] Indeed, they use their signals to help target specific bodyguards to fend off different types of assailants. When a caterpillar munches on a leaf, the tree releases one set of terpenes. When a beetle burrows past the bark to eat the tree marrow (phloem), another terpene is released. Trees summon their specialized army of insect defenders at will, which is why they can stay healthy despite not being able to pluck or crush the bugs that bother them. Trees even use terpenes to control air temperature. On hot summer days, more terpenes are released than on cold days. As they rise into the atmosphere above the forest, they seed the atmosphere and create clouds, helping to block sunlight and cool the forest, like an on-demand air conditioning system. And that's not all: trees also release chemicals to attract pollinators, fight infections (such as fungal growths), and even warn neighboring trees of pest invasions. Each tree produces a unique pharmacy, dispensing specific medicines when needed.

Terpenes are not only useful for trees; they are also useful for us. When we walk through a pine forest, we inhale terpenes—biochemicals that have antibacterial properties. Some research shows they are effective in combating respiratory tract pathogens like *E. coli* and can even be used to treat breast cancer. Some scientists argue that terpenes might even be neuroprotective. If so, breathing in the smell of real pine trees may be, quite literally, medicinal. As Robin Wall Kimmerer writes: "Plants know how to make food and medicine from light and water, and then they give it away."[26] Environmentalists opposed to VR and AR assert that this is something for which virtual reality and augmented reality can't substitute. Or at least not yet: spritzing chemicals like terpenes during VR experiences may be on offer in the near future.

The broader point is that digital nature is no substitute for the real thing: humans are hardwired to need a direct physical experience of the Earth. As Abram puts it in *The Spell of the Sensuous*, "we don't live *on* the Earth. We live *in* the Earth." Even when wearing a VR headset, we breathe in oxygen from the Earth's atmosphere. Because air is invisible, we often treat it as nonexistent (or treat it like an open sewer), but air is something tangible that connects us to other beings on Earth: they breathe out, we breathe in. Evangelista Torricelli (a student of Galileo best known for inventing the barometer) famously wrote in 1644, "We live submerged at the bottom of an ocean of . . . air."[27] To live is to breathe and to metabolically exchange with others on the planet. For Abram, this has a spiritual dimension, reflected in etymology: our word for atmosphere comes from the Greek *atmos*, or vapor, which in turn is derived from the Sanskrit word for breath, *atman*, which is also Hindu for soul, essence. Digital technologies interrupt rather than enhance our sensual, vibrant, spiritual connection to our shared biophysical connection as living beings.

LOVING NATURE . . . BUT NOT TO DEATH

Although virtual or augmented reality experiences can never be fully equivalent to actual face-to-place experiences with nature, they are nonetheless something that environmentalists should consider embracing. They are indeed imperfect versions of the real thing, but they still expand our knowledge of nature. I'm not likely to ever climb Mount Everest in real life but would be keen to try climbing the world's highest peak in an augmented reality scenario. I'm equally unlikely to visit the depths of the ocean, but I'd love to have a VR experience that takes me there.

VR/AR could enable our increasingly urbanized population to access nature experiences more frequently than is currently possible, and perhaps reduce our environmental footprint. Over half of the world's population is now urban, and the rate of urbanization is predicted to continue accelerating in the future. Without something like virtual and augmented reality, most of the world's population will become ever more distanced from nature. And by providing these experiences, we could dramatically lessen the impact of actual humans on actual nature, while allowing more people to access nature virtually.

What if even 1 percent of humans on the planet decided that they needed a real-life experience in Yellowstone this year? The remaining pockets of wilderness on Earth can't physically accommodate that many people having direct experiences of nature, who might trample it to death. Ontario farmer Brad Bogle can attest to this firsthand. One midsummer day in 2018, he sent out an invitation on social media for people to come and take selfies in his field of 1.4 million sunflowers on July 28, the peak time for sunflower blooms. Unexpectedly, the invitation went viral. By the time "Crazy Day" (as Bogle later began calling it) rolled around, an estimated 7,000 people visited the farm. By mid-morning, Bogle was turning people away and had put up "no trespassing" signs, but many visitors refused to comply—swearing at Bogle and his family, sneaking onto the property, driving through fields, getting into fender benders. One man even threatened Bogle with a fistfight when asked to leave. Families began parking up to a mile down the shoulderless highway, crossing four lanes of traffic with strollers, and marching to the property, a few urinating in neighbors' yards along the way. One driver had his door ripped off by a passing car. Even emergency police blockades didn't stop the zombie apocalypse of sunflower-hungry urbanites.

Maybe we should thank our lucky stars that most people aren't interested in having direct experiences with nature. Many delicate ecosystems have suffered from the effects of masses of overeager tourists. The ramping up of mass tourism in US parks like Yosemite has led to traffic jams and the killing of black bears who could become dangerous "nuisances" to visitors. India's most famous lake—Dal Lake, nestled in the foothills of the Himalayas—is now choked with weeds and sewage. The effects of tourism on delicate Arctic moss in Iceland are so devastating that the phenomenon now has a scientific name: "recreational trampling." With the recent surge of tourism to the island, carving selfie-friendly messages in the ancient moss became a viral phenomenon; scientists say that it will take centuries to recover from what they term "nature vandalism." As COVID-19 changed travel patterns in 2020, similar stories emerged about "Rivergeddon" in Montana, where the Madison River was so crowded with boats that nature enthusiasts lamented the deluge of tourists, creating an even busier outdoor version of Disney's Pirates of the Caribbean ride.

Although almost woefully ironic as a solution, virtual and augmented reality may end up saving what remains of nonhuman nature from humanity. Humans are now the dominant species on Earth, and highly efficient at harvesting the biosphere. To put this in perspective, the combined weight of all wild land mammals is only about a tenth of the combined weight of the world's human population.[28] Humanity has already eaten or exterminated most wild nonhuman creatures on Earth. And given predictions of demographic growth—the Earth's human population will grow by an estimated 1.7 billion in the next thirty years—perhaps environmentalists should be thankful that technologies now exist that allow us to love nature at a safe distance. Pitting "real" against "fake" nature may also be a false paradox, or simply an outdated one. Increasingly, we can augment and enhance our experience of nature with technologies that enhance our appreciation and sense of wonder. Hundreds of smartphone apps identify plant species (FlowerChecker, NatureGate, PlantSnap), landscape features (ViewRanger), and animals (Merlin, iNaturalist).[29]

Future innovations could enable us to walk through a forest and see species labeled in real time simply by holding up our cell phones, much like the virtual labels being used in retail shops. In a world where much of the knowledge of the natural world has been lost, augmenting our experiences of nature in this way could help reeducate us about nature. Image recognition algorithms can also be applied to plants—for example, as in LeafSnap, which automatically identifies tree species from photos of their leaves. Enrolling citizen scientists is another potential benefit: smartphones are helping to build enormous repositories of observations (such as eBird), while engaging new communities (such as tourists) in conservation. The next generation of these apps will also include contextual data; computer vision will not only focus on naming/identifying the species but also assess the surroundings for species identification and biodiversity—just as a "real" naturalist would.

ACORN, BLUEBELL, CONKER . . . ANALOG, BROADBAND, CELEBRITY

Virtual, augmented, and extended reality can also help us rediscover that we, too, are part of nature. Modernity is predicated on dominating plants and animals while forgetting that we ourselves are animals. To become modern, our culture developed a generalized amnesia about this fact and, as with all repressions, a collective anxiety about our amnesia. Our children retrace this historical shift as they grow up: young children's innate affiliation with nature is universal and intense. But in highly technologically mediated societies, children's biophilia (love of nature) declines as they become teenagers.[30]

By the time they are adults, most have lost interest in the natural world. And this phenomenon is accelerating: children's vocabulary has fewer nature words than in the past. The most recent version of the *Oxford Junior Dictionary* (aimed at kids seven and up) has included fewer and fewer nature words in recent editions. Acorn, buttercup, conker: make way for analog, broadband, and celebrity. Instead of words like "catkin," "chestnut," and "clover," the dictionary features new terms like "cut and paste." For blackberry, read "BlackBerry." These changes in language reflect broader shifts in the patterns of our attention, particularly for children, many of whom spend more time on screens than outdoors.

Virtual and augmented reality technologies could help us reverse this distancing from nature. At least, that is the hope of innovators like Jeremy Bailenson. But what if these technologies do the opposite of what Bailenson intends? What about someone who feels that VR is an adequate substitute for nature? Is there a risk that they would feel fine about chopping down our last remaining old-growth forests, and simply be content with virtual forests instead? Could these technologies make us less worried about biodiversity loss because we can recreate nature virtually? Some examples—like internet hunting, which enables users of remotely controlled firearms to shoot and kill animals via webcams—foreshadow a dystopian future.

Which path we follow depends in part on how well we avoid a common pitfall: confusing cognitive and emotional empathy. Cognitive empathy means viewing something from another perspective. Emotional empathy means feeling someone's emotions. These two types of empathy don't always go together. Some bullies have high cognitive empathy—which helps them better understand and manipulate their victims—but low (or "reversed") emotional empathy, which enables them to persist in hurting others. If you experience only cognitive empathy, you might understand how to act in a positive way but feel unmotivated to do so. Most psychologists agree that a combination of cognitive and emotional empathy is required for people to change their behaviors constructively. So, VR and AR are likely to trigger us to conserve and protect the environment only to the extent that they trigger both cognition and emotion. We need to move from an emotional response to rational compassion.[31]

Yet emotion and reason, even if necessary, aren't sufficient to get us to act. For that, a third ingredient is required. Empathy also requires some "fellow feeling" for others, which is usually rooted in shared values or feelings of similarity or relatability. When thinking about environmental conservation, this seems unrealistic at first glance. A "fellow feeling" is sometimes hard enough to muster for our fellow humans. In fact, some studies suggest that human empathy has declined over the past few decades (and, alarmingly, the rate of decline may be increasing).[32] This is compounded by the fact that humans tend to find it easier to feel empathy for single individuals in pain rather than for large numbers of suffering masses. A single child dying of starvation moves us to action (or at least donations). Much attention has been showered on the 52 Hz whale—also known as the "loneliest whale in the world"—which communicates at a higher frequency than others of its kind, limiting its socialization. But the fate of the whales killed for scientific experimentation every year arouses less interest. This is why environmental organizations tend to focus on single stories: while reading about the plight of flocks of birds in an oil spill might leave us cold, we are more likely to respond when we read about one suffering baby loon.

There is one drawback with this individual story–driven approach: most environmental crises don't have charismatic megafauna readily available. Instead, they often involve large numbers of anonymous creatures. Millions of corals dying after Australia's heat wave of 2016 didn't incite mass action. If most people don't even feel much empathy for their neighbors, how can they be expected to feel empathy for, say, a school of fish? The answer to this question may lie in a fourth key ingredient in empathy: social affiliation, or a feeling of belonging. We traditionally think of belonging to different human communities: family, tribe, class, or nation. But our need for belonging is not bounded by humankind. By helping us cultivate a sense of fellow feeling with nonhumans, augmented reality technologies could extend our "circle of sympathy," as Steven Pinker puts it. Skeptics argue, however, that empathy is difficult enough to achieve between members of our own species, and there is no guarantee that humans will easily extend it to other species. In opposition to calls for greater "conviviality" with nature, critics insist on the need for a more old-fashioned but proven approach: strict environmental protection and conservation mechanisms—like parks and protected areas. Rather than hoping that humans might develop greater empathy through greater exposure to nature, a better strategy would be to protect nature, as much of it as possible, from too much human interaction. The hopes of VR/AR innovators are, from this perspective, misguided.

THE HAPTIC INTERNET

In addition to developing new innovations in VR and AR, digital innovators have also been exploring the potential uses of haptic—touch-based—technologies to enable humans to connect with animals. For instance, computer scientist Adrian Cheok's "Internet of Poultry" experiments enable users to remotely hug chickens via a Wi-Fi–enabled haptic vest. Cheok has also invented devices that enable you to remotely pat your domestic pets, and even kiss your distant lovers (via Kissenger, a multisensory real-time kissing app for your mobile phone). Cheok imagines a "huggable internet," which digitally enables humans, robots, and other species to interact.

Other haptic innovations are less gimmicky.[33] Take, for example, the Cyber-Enhanced Working Dog harness: this cyber-enabled, computer-mediated communications system provides both human and canine partners with real-time information about each other's behavioral and emotional states through postural/movement detection, heart rate and respiration monitoring, and haptic feedback and auditory cues. For now, this technology is focused on enhancing the performance of seeing-eye dogs and search-and-rescue dogs. But in the future, when combined with robotic and autonomous technologies, interactive communications systems like cyber harnesses will form a core part of the basis of human-animal symbiont emergency response systems. Other applications are obvious: companion animals for mental health and palliative care patients, and even enhancing our relationships with our pets. It's not so difficult to imagine a digitally enabled interface for interacting with companion animals—after all, we touch and stroke them all the time. But haptic technologies extend the possibility of interaction to potentially all living things (Table 8.1). Could they bridge the gap between Abram's invocation of embodied, earthly sensuality and the

Table 8.1
Haptic animal-human interaction technologies

Type of technology	Species	Example
Wearables	Pet dogs	FitBark dog activity wearable monitor[a]
	Pet cats	Commercial cat wearables such as Pawtrack that use GPS collars incorporating accelerometers and cameras;[b] assistive wearables that support deaf-blind cat owners[c]
	Pigs	iPig: wearable sensor for detecting domesticated pig movements[d]
	Cows	Cow wearables that detect gait anomalies and other health-related metrics[e]
Interactive touchscreens	Pigs	Interactive touch screen that enables domesticated pigs and humans to play an interactive game (Pig Chase)[f]
	Cats	Interspecies tablet game for cats[g]
	Orangutans	Large-scale zoo interface via Microsoft Kinect sensor and projector[h]
	Dogs	Touchscreen interface for working dogs[i]
	Dolphins	Underwater touchscreen keyboards[j]
	Elephants	Vibrotactile buttons designed to provide haptic feedback to elephants via their trunks[k]
	Crickets	Vibrotactile maze that provides feedback to crickets playing Pac-Man with human players[l]
	Chickens	Chicken wearable and mixed reality system for human-poultry interaction[m]

a. "Dog GPS & Health Trackers," FitBark, accessed May 8, 2023, https://www.fitbark.com/.
b. Kyoko Yonezawa, Takashi Miyaki, and Jun Rekimoto, "Cat@ Log: Sensing Device Attachable to Pet Cats for Supporting Human-Pet Interaction," in *Proceedings of the International Conference on Advances in Computer Entertainment Technology*, Athens, Greece, October 29–31 (New York: Association for Computing Machinery, 2009), 149–156.
c. Alexander König, "Designing an Assistive Wearable That Supports Deafblind Cat Owners," master's thesis, Malmö University, 2020.
d. Juan Haladjian, Ayca Ermis, Zardosht Hodaie, and Bernd Brügge, "iPig: Towards Tracking the Behavior of Free-Roaming Pigs," in *Proceedings of the Fourth International Conference on Animal-Computer Interaction*, Milton Keynes, United Kingdom, November 21–23 (New York: Association for Computing Machinery, 2017), 1–5.
e. Juan Haladjian, Zardosht Hodaie, Stefan Nüske, and Bernd Brügge, "Gait Anomaly Detection in Dairy Cattle," in *Proceedings of the Fourth International Conference on Animal-Computer Interaction*, Milton Keynes, United Kingdom, November 21–23, 1–8 (New York: Association for Computing Machinery, 2017); Francisco Carpio, Admela Jukan, Ana Isabel Martín Sanchez, Nina Amla, and Nicole Kemper, "Beyond Production Indicators: A Novel Smart Farming Application and System for Animal Welfare," in *Proceedings of the Fourth International Conference on Animal-Computer Interaction*, Milton Keynes, United Kingdom, November 21–23 (New York: Association for Computing Machinery, 2017), 1–11; Suresh Neethirajan, "Recent Advances in Wearable Sensors for Animal Health Management," *Sensing and Bio-Sensing Research* 12 (2017): 15–29.
f. Mikhail Fiadotau, "Digital Games for Animals," in *Encyclopedia of Computer Graphics and Games*, ed. Newton Lee (Cham: Springer, 2018).

Table 8.1 (continued)

g. Patricia Pons and Javier Jaen, "Towards the Creation of Interspecies Digital Games: An Observational Study on Cats' Interest in Interactive Technologies," in *Proceedings of the 2016 CHI Conference Extended Abstracts on Human Factors in Computing Systems*, San Jose, California, May 7–12 (New York: Association for Computing Machinery, 2016), 1737–1743; Michelle Westerlaken and Stefano Gualeni, "Felino: The Philosophical Practice of Making an Interspecies Videogame," in *Proceedings of the Philosophy of Computer Games*, Istanbul, Turkey, November 13–15 (Istanbul: Istanbul Bilgi University, 2014), 1–12.
h. Marcus Carter, Sally Sherwen, and Sarah Webber, "An Evaluation of Interactive Projections as Digital Enrichment for Orangutans," *Zoo Biology* 40, no. 2 (2021): 107–114; Hanna Wirman, "Orangutan Play on and beyond a Touchscreen," in *Proceedings of the 19th International Symposium on Electronic Art*, Sydney, Australia, June 7–16 (Adelaide: ISEA International Australian Network for Art & Technology, 2013), 1–3.
i. Clint Zeagler, Scott Gilliland, Larry Freil, Thad Starner, and Melody Jackson, "Going to the Dogs: Towards an Interactive Touchscreen Interface for Working Dogs," in *Proceedings of the 27th Annual ACM Symposium on User Interface Software and Technology*, Honolulu, Hawaii, October 5–8 (New York: Association for Computing Machinery, 2014), 497–507; Clint Zeagler, Jay Zuerndorfer, Andrea Lau, Larry Freil, Scott Gilliland, Thad Starner, and Melody Moore Jackson, "Canine Computer Interaction: Towards Designing a Touchscreen Interface for Working Dogs," in *Proceedings of the Third International Conference on Animal-Computer Interaction*, Milton Keynes, United Kingdom, November 15–17 (New York: Association for Computing Machinery, 2016), 1–5.
j. Denise L. Herzing, "Interfaces and Keyboards for Human-Dolphin Communication: What Have We Learned?," *Animal Behavior and Cognition* 3, no. 4 (2016): 243–254.
k. Fiona French, Clara Mancini, and Helen Sharp, "High Tech Cognitive and Acoustic Enrichment for Captive Elephants," *Journal of Neuroscience Methods* 300 (2018): 173–183.
l. Wim Van Eck and Maarten H. Lamers, "Animal Controlled Computer Games: Playing Pac-Man against Real Crickets," in *Entertainment Computing—ICEC 2006*, ed. Richard Harper, Matthias Rauterberg, and Marco Combetto (Berlin: Springer, 2006), 31–36; Wim Van Eck and Maarten H. Lamers, "Player Expectations of Animal Incorporated Computer Games," in *Proceedings of the International Conference on Intelligent Technologies for Interactive Entertainment*, Funchtal, Portugal, June 22–25 (Cham: Springer, 2017), 1–15.
m. Shang Ping Lee, Adrian David Cheok, James Teh Keng Soon, Goh Pae Lyn Debra, Chio Wen Jie, Wang Chuang, and Farzam Farbiz, "A Mobile Pet Wearable Computer and Mixed Reality System for Human-Poultry Interaction through the Internet," *Personal and Ubiquitous Computing* 10, no. 5 (2006): 301–317.

digital spell of the Magicverse? Or would they merely deepen the illusion, obscuring rather than enhancing our connection to living things and places?

ASK THE RIVER

Immersive reality designers hope to make us to feel what animal biologists call the *umwelt* (literally, the "world around") of nonhumans. In the process, they seek our understandings not only of other species but also of ourselves. Yet contrary to the claims of its boosters, increased empathy as a result of immersive technology experiences is often at best a transient phenomenon, and at worst a mirage. Philosopher Liora Gubkin advocates a different approach: replacing empathetic understanding with "engaged witnessing" as a framework for teaching about traumatic knowledge.[34] Contemporary environmentalism is indeed witnessing a collective trauma: the colonization of Gaia by industrial modernity, and the creation of the technosphere, requiring enormous amounts of energy to harvest resources and colonize land and water—resulting in the climate change and biodiversity crises we are experiencing today.

Indigenous artists offer another approach to immersive reality technologies, in which engaged witnessing of relationships to land is linked to an assertion of resistance and reimagining reciprocal relationships to land and place. What if virtual and augmented reality projects were "spaces to be navigated through," rather than merely films to be experienced? Keziah Wallis, a Māori anthropologist from New Zealand, argues that Indigenous futurisms provide a more grounded basis for environmental action. In her curated database of Indigenous VR (fourthVR.com), Wallis and her collaborators explore how VR creates spaces of decolonization, in which native languages are foregrounded, Indigenous activism is articulated, and the interconnectivity of all living things—past, present, and future—is demonstrated.[35]

In *Thalu: Dreamtime Is Now*, Ngarluma artist Tyson Mowarin takes the viewer into northwestern Australia, beginning with a story of Earth's creation, then plunging the viewer into a kinesthetic experience of place in the spirit world that is the essence of the land. In spoken Ngarluma, the user is greeted as a steward of the land and tasked with being a protector—which requires learning the laws of the land. The user is guided through lessons about caring for the country—creating rain, gifting plants and animals to the human world, and protecting sacred sites, which in turn protect Earth's bounty and abundance. Unsettling the settler, while offering another way of living in relationship to the land, *Thalu* is an experience in which time and space collapse while the terrestrial world expands to the spirit world, which Mowarin describes as alive and ever-present. Unlike conventional VR, *Thalu* offers an experience that fuses land and body, future and memory.

Indigenous VR is both a futuristic technology and a mechanism to remember and question settler colonial narratives about the past. As Cree/Métis filmmaker Loretta Todd writes: "It is not so odd, then, at this stage of late capitalism in the project called western culture, that cyberspace is under construction. It has in fact been under construction for at least the past two thousand years in western cultures. A fear of the body, an aversion to nature, a desire for salvation and transcendence from this earthly plane has created a need for cyberspace. The wealth of the land almost plundered, the air dense with waste, the water sick with poisons: there has to be somewhere else to go."[36] Cyberspace, as conventionally designed, is an elaborate metaphor that seeks alienation and separation under the guise of observation. As Todd notes, the experience of nature in Indigenous VR expresses a worldview of "subjects to subjects, consciousness to consciousness," in which human bodies are connected to the material world. As Kanien'kehaka multimedia artist Jackson 2Bears asks about the VR depiction of a Coast Salish longhouse by Cowichan/Syilx artist Yuxweluptun: Can we think about VR helmets as "being analogous to the *Sxwaixwe* mask, a spiritual mediator between the incommensurable, death and life, embodiment and disembodiment, virtuality and flesh? That is, considered as a reversal of the codes of simulation, can technology here become hauntological, where dreams and visions are synonymous with that of technological immersion at the site of the collapse between the boundaries of the virtual and biological organisms?"[37]

Hauntology is the notion that the present is haunted by futures foregone, as well as the past; it points to an uncanny experience of time as circular and cyclical rather than linear. As time folds, past and future experiences enter into the present moment. Immersive reality reminds us that there are other ways of listening, of seeing, and of being as time collapses into a singular moment. At that site of collapse—figurative and literal—the land appears as a character in the narrative, a nonhuman individual with enduring memories, relationships, and teachings. Nonhumans appear as subjects rather than objects: nonhuman individuals and nonhuman kin. As the next chapter explores, this prefigures the hybrid technologies now emerging, in which biological and digital entities are being fused, merged, and hybridized—with startling implications for environmental governance and for our understanding of what it means to be human in a digital age.

PARABLE OF MOTHBOT

MothBot flutters on the night wind, following the light of the moon. Thousands of miles away, a technician updates some code, and MothBot slows and turns.

Before she was born, while still in her cocoon in a lab in North Carolina, two electrodes were implanted in her ventral nerve. As her pupa developed, her ventral nerve cord fused to the wires, creating a seamless bond with its living tissues. MothBot has never perceived the electrodes as foreign objects. She came into this world as a living drone, engineered to be controllable at a distance.

Exquisitely sensitive to smell, she can detect chemicals commonly found in explosives. The code directs her to a nearby minefield, sensing hot spots and safe zones, a silent sentinel.

When she is longer useful, she will be discarded.

Thousands of miles away, her mother begins her nightly rounds, checking the factory production line that rolls out thousands of MothBots a day.

When her coworkers are not looking, she sings softly to her children.

9 BIOCYBORG KIN

In a small lab in Raleigh, North Carolina, a graduate student is working late into the night. A tiny cocoon the size of a pencil eraser sits on her workstation: a pupal moth, yet to be born. Bent over the cocoon, she handles her instruments carefully. Her job is to implant tiny, flexible microprobes into a precise spot on the pupa inside the cocoon, without damaging it. The technique she is using—Early Metamorphosis Insertion Technology—takes advantage of the uniquely receptive nature of the fast-growing immature moth.[1] As the pupa develops, its ventral nerve cord will fuse to the implanted wires, ensuring a seamless bond with its living tissues. When it emerges, the insect will not perceive the electrodes as foreign objects. The student will stimulate the nerve with small electrical pulses, either through a wire tether or remotely via a radio signal.[2] When she stimulates the nerve, the moth's muscles will contract, which will change its direction as it flies. The MothBot will be controllable at a distance, a living drone.

In another lab, in Cleveland, Ohio, a graduate student is creating a SlugBot.[3] First, she cultures a muscle from the mouth of a living slug. Next, she fuses the soft, flexible tissue—about the size of a newborn's pinky finger—to a 3D plastic printed scaffold, which is made of a biocompatible synthetic polymer. The resulting creature looks like a hapless inchworm perched atop a dental retainer. Instead of synthetic wires, the slug's living muscle and nerve cells will be used as actuators. Completely waterproof, the SlugBot can carry sensors as it moves, in an eerie parody of its real-life cousin.

Another biohybrid robot, made using a Madagascar hissing cockroach, holds particular promise for emergency response. The roach, commonly referred to as the Hisser, is one of the largest species of cockroaches in the world; adults are commonly two to three inches long. Their relatively large backs provide a perfect platform for digital technology, including biofuel cells that can power miniaturized digital devices. Equipped with wireless electronic backpacks, which are powered by miniaturized solar panels, and implanted with electrodes, multiple roaches can be controlled and directed simultaneously to scuttle in systematic patterns within a defined perimeter, scanning a search zone. Researchers hope that these digitally enhanced roaches will soon be used to crawl under the rubble of collapsed buildings, through the tiniest of crevices, to locate earthquake survivors.[4]

MothBot, SlugBot, and RoachBot are just three additions to our rapidly expanding pantheon of animal-machine hybrids which scientists refer to as biobots, or biohybrid robots.[5] In the past few years, living, breathing, creeping, crawling biobots have been created with cells harvested from beetles, ray fish, fall armyworms, giant flower beetles, California sea slugs, jellyfish, zebrafish, tobacco hawk moths, turtles, honeybees, rats, and dragonflies (see table 9.1). Some biobots are life-sized. Others are microscopic, such as the sperm-like swimming biobots that use cardiac muscle cells, cultured from rodents, to propel themselves forward. In some cases, control is automated through a rudimentary neural-computer interface, with digital probes stimulating specific areas of the animal's brain.[6]

In this chapter, we will explore the uses of biobots as environmental sentinels and monitoring devices. The development of these devices promises to enhance environmental monitoring and remediation. However, the proliferation of biological robots also raises serious ethical questions about the future evolution of humanity as a species, and humanity's relationship with nature. Recent advances in gene editing technology (e.g., CRISPR), artificial intelligence, materials technology, and other technologies have radically altered our ability to manipulate life and fuse living organisms with machines in unprecedented ways, yet few ethical guideposts, and guardrails, exist in this strange new world.

Table 9.1
Environmental applications of biobots

Biobot	Method	Proposed application
Electrified snail	An implanted biofuel cell continuously operates in the snail, producing electrical power over a long period of time using the snail's own naturally produced glucose as fuel. The "electrified" snail, being a biotechnological living "device," can regenerate glucose consumed by the biofuel cell; feeding and resting enable it to produce a new supply of electrical energy.[a]	Electrified snails with implanted biofuel cells will be able to operate as stand-alone sensors in natural environments while producing sustainable electrical micropower for activating bioelectronic devices.
Remote-controlled American grasshopper	Micro-electrodes implanted directly into the insect's brain were used to stimulate, with constant electrical pulses, the metathoracic T3 ganglion, which coordinates the neuromuscular activity leading to a jump.[b]	Remote-controlled cyborg insects that will jump on command have future uses as micro air vehicles (MAVs) that can be used in military applications, environmental sensing, and search-and-rescue missions.
Beetle with cyborg eyes	Flexible neural interfaces are inserted in eyes of *Zophobas morio* Fabricius (Coleoptera: Tenebrionidae) during the pupal stage. During metamorphosis, projections that extend from the neurons in the insect's sensory organ grow through the implant, resulting in an adult insect with a stable and fused organ-(artificial) nervous system.[c]	Eye implants in beetles create "cyborg eyes" that can gather neural recordings and transmits them to remote devices.
Light-controlled manta ray	A biohybrid system enables an artificial animal—a tissue-engineered ray—to swim and move away from, or toward, a light cue.[d]	Biohybrid sea creatures can be guided by remote optical signaling.
Worm-cell augmented chemical sensor	Cells from the fall armyworm were utilized in the creation of an electronic odorant sensor to discriminate between two similar odorant responses with high reliability.[e]	Cyborg worms can be used to detect chemicals via odor discrimination, without the need for other chemical testing.
Biohybrid sea creature scaffold for microelectronics	Isolated collagen from California sea hare skin is used to create scaffold material for the 3D-printed structure of the biohybrid robot, while isolated muscle tissue is used as an organic actuator. These aspects create a cyborg capable of muscle strain and force, and locomotion.[f]	The marine cyborg is entirely biodegradable with an entirely customizable scaffold for microelectronic devices.
Remote-controlled jellyfish	A self-contained microelectronic swim controller is embedded into a jellyfish's tissue, which then generates pulses to stimulate muscle contractions.[g]	Jellyfish physical performance is augmented, tripling swimming speeds and doubling metabolic expenditure.

Table 9.1 (continued)

Biobot	Method	Proposed application
Autonomous communication of injury by a plant to a smartphone	Single-wave carbon nanotube sensors were incorporated into plant leaves of mouse-ear cress plants. When the plants were injured, these sensors detected wound-induced hydrogen peroxide. The nanotubes can also be used to detect other stressors in plants, like high heat and intense lighting.[h]	Nanotube-augmented signals can be picked up by a light sensor, and the results communicated to an external internet-enabled device.
Living cockroach biobattery	A female cockroach is used to create a living biobattery by implanting the insect with a biofuel cell that uses trehalose, a kind of glucose that naturally occurs in the insect, and the oxygen in the air around it to generate electricity that can be collected and stored for later use.[i]	Insects can use energy from their bodies to recharge biobatteries, which can be used to power portable electronic sensors.
Radio-controlled beetle	An electrical neural interface enables radio control of a live giant flower beetle's flight, prompting it to "go," "stop," and "turn." This is accomplished using a microcontroller and six electrode stimulators implanted into the brain and flight muscles.[j]	Remote control of insect-free flight can be demonstrated in the giant flower beetle.
Self-powered, remote-controlled search-and-rescue Madagascar hissing cockroach	A robotic backpack outfitted with a Microsoft Kinect camera and three unidirectional microphones (designed as an acoustic sensor) is fitted onto the cockroach, which then acts as an autonomous sensor for environmental mapping and search-and-rescue missions. In other experiments, electrodes are surgically implanted into the antenna of the cockroach to enable remote control of its direction and movement.[k]	Remote-controlled insects can be used as self-powered, remote-controlled search-and-rescue devices.
Radio-controlled tobacco hawk moth	Micro-metal wire stimulation probes are implanted into an adolescent moth's brain and thoracic tissue, enabling the robotic probes to be accepted and grow as part of the moth's body during metamorphosis. After this stage, a helium balloon is attached via magnet to the probe to offset the added weight. Through remote radio signals, observers were able to control the flight path of the moth.[l]	Insect flight can be remotely radio-controlled via an electrical neural interface.

Table 9.1 (continued)

a. Lenka Halámková, Jan Halámek, Vera Bocharova, Alon Szczupak, Lital Alfonta, and Evgeny Katz, "Implanted Biofuel Cell Operating in a Living Snail," *Journal of the American Chemical Society* 134, no. 11 (2012): 5040–5043; Evgeny Katz and Kevin MacVittie, "Implanted Biofuel Cells Operating in Vivo: Methods, Applications and Perspectives—Feature Article," *Energy & Environmental Science* 6, no. 10 (2013): 2791–2803.
b. Susan L. Giampalmo, Benjamin F. Absher, W. Tucker Bourne, Lida E. Steves, Vassil V. Vodenski, Peter M. O'Donnell, and Jonathan C. Erickson, "Generation of Complex Motor Patterns in American Grasshopper via Current-Controlled Thoracic Electrical Interfacing," in *Proceedings of the 2011 Annual International Conference of the IEEE Engineering in Medicine and Biology Society*, Boston, August 30–September 3 (New York: Institute of Electrical and Electronics Engineers, 2012), 1275–1278.
c. Amol D. Jadhav, Ivan Aimo, Daniel Cohen, Peter Ledochowitsch, and Michel M. Maharbiz, "Cyborg Eyes: Microfabricated Neural Interfaces Implanted during the Development of Insect Sensory Organs Produce Stable Neurorecordings in the Adult," in *Proceedings of the 2012 IEEE 25th International Conference on Micro Electro Mechanical Systems (MEMS)*, Paris, France, January 29–February 2 (New York: Institute of Electrical and Electronics, 2012), 937–940.
d. Sung-Jin Park, Mattia Gazzola, Kyung Soo Park, Shirley Park, Valentina Di Santo, Erin L. Blevins, Johan U. Lind, et al., "Phototactic Guidance of a Tissue-Engineered Soft-Robotic Ray," *Science* 353, no. 6295 (2016): 158–162.
e. Daigo Terutsuki, Hidefumi Mitsuno, Yuki Okamoto, Takeshi Sakurai, Agnès Tixier-Mita, Hiroshi Toshiyoshi, Yoshio Mita, and Ryohei Kanzaki, "Odor-Sensitive Field Effect Transistor (OSFET) Based on Insect Cells Expressing Insect Odorant Receptors," in *2017 IEEE 30th International Conference on Micro Electro Mechanical Systems (MEMS)*, Las Vegas, Nevada, January 22–26 (New York: Institute of Electrical and Electronics Engineers, 2017), 394–397.
f. Victoria A. Webster, Katherine J. Chapin, Emma L. Hawley, Jill M. Patel, Ozan Akkus, Hillel J. Chiel, and Roger D. Quinn, "*Aplysia californica* as a Novel Source of Material for Biohybrid Robots and Organic Machines," in *Biomimetic and Biohybrid Systems: 5th International Conference, Living Machines 2016, Edinburgh, UK, July 19–22, 2016. Proceedings*, ed. Nathan F. Lepora, Anna Mura, Michael Mangan, Paul F. M. J. Verschure, Marc Desmulliez, and Tony J. Prescott (Cham: Springer, 2016), 365–374.
g. Nicole W. Xu and John O. Dabiri, "Low-Power Microelectronics Embedded in Live Jellyfish Enhance Propulsion," *Science Advances* 6, no. 5 (2020): 1–10.
h. Tedrick Thomas Salim Lew et al., "The Emergence of Plant Nanobionics and Living Plants as Technology," *Advanced Materials Technologies* 5, no. 3 (2020): 1–12; Tedrick Thomas Salim Lew et al., "Real-Time Detection of Wound-Induced H_2O_2 Signalling Waves in Plants with Optical Nanosensors," *Nature Plants* 6, no. 4 (2020): 404–415.
i. Michelle Rasmussen, "Trehalose-Based Biofuel Cells," PhD diss., Case Western Reserve University, 2012.
j. Hirotaka Sato, Yoav Peeri, Emen Baghoomian, Christopher W. Berry, and Michel M. Maharbiz, "Radio-Controlled Cyborg Beetles: A Radio-Frequency System for Insect Neural Flight Control," in *2009 IEEE 22nd International Conference on Micro Electro Mechanical Systems*, Sorrento, Italy, January 25–29 (New York: Institute of Electrical and Electronics Engineers, 2009), 216–219.
k. Alper Bozkurt, Edgar Lobaton, Mihail Sichitiu, Tyson Hedrick, Tahmid Latif, Alireza Dirafzoon, Eric Whitmire, et al., "Biobotic Insect Swarm Based Sensor Networks for Search and Rescue," in *Signal Processing, Sensor/Information Fusion, and Target Recognition XXIII vol. 9091*, Baltimore, MD, May 5–9, ed. Ivan Kadar (Bellingham, WA: International Society for Optics and Photonics, 2014), 498–503; Tahmid Latif, Eric Whitmire, Tristan Novak, and Alper Bozkurt, "Sound Localization Sensors for Search and Rescue Biobots," *IEEE Sensors Journal* 16, no. 10 (2015): 3444–3453; Eric Whitmire, Tahmid Latif, and Alper Bozkurt, "Acoustic Sensors for Biobotic Search and Rescue," in *Proceedings of SENSORS*, Valencia, Spain, November 2–5, 2014 (New York: Institute of Electrical and Electronics Engineers, 2014), 2195–2198.
l. Alper Bozkurt, Amit Lal, and Robert Gilmour, "Radio Control of Insects for Biobotic Domestication," in *2009 4th International IEEE/EMBS Conference on Neural Engineering*, Antalya, Turkey, February 23–March 13 (New York: Institute of Electrical and Electronics Engineers, 2009), 215–218.

THE BIOBOTS ARE HERE

Scientists have long dreamt of transforming living creatures into remotely controlled biobots by fusing them with engineered components. Until recently, such visions remained confined to the realm of science fiction, but in the past five years, a proliferation of biobots has been enabled by advances in biofabrication, nanotechnology, neurocybernetics, and bionics. Researchers are now able to transform almost any living creature into a biobot by fusing embryos or cultured living tissue (such as cardiac or skeletal muscle cells) with electronic components designed as nerve and muscle cell stimulators. With microscopic precision, these hybrid devices are hooked up to sensors, computers, an array of remote-control devices, and smartphones, enabling precise control of biobots at a distance. These innovations are examples of what is known as *biodigital convergence*: the intermingling of biological and digital innovation, in which digital technologies are increasingly biological, and biological organisms are increasingly digital.

Some biobot researchers aspire to help humans through medical innovations in prosthetics or self-healing materials. A burgeoning field of inquiry in bioengineering has produced a myriad of promising inventions. Researchers have, for example, created biological SpinoBots, which fuse 3D-printed mouse muscle cells to the section of a rat's spinal cord that controls the hind legs, allowing the researchers to operate the rat's legs with neurotransmitters. Eventually, its designers hope, this type of technology could be used in prosthetics or could be implanted in human bodies.

Biobots also have important implications for Digital Earth governance, as both environmental monitoring devices and vectors for environmental conservation interventions. SlugBots, which are completely biodegradable, could be used as environmental sensors; at the end of their useful life as monitoring devices, they would simply decompose wherever they are left to die. Much like sniffer dogs, MothBots could be used to detect dangerous chemicals and explosives; they could also be used to map ecosystem biodiversity.

Swarms of biobot bats could vastly accelerate environmental monitoring or help locate survivors after natural disasters.[7] Swarm robots—autonomous, nonsynchronized, nonintelligent micro-robots that cooperate to achieve tasks—are revolutionizing other areas of environmental conservation as well (table 9.1). Larva Bots, for example, roam coral reefs and shoot out baby coral in an attempt to slow down the damage caused by climate change. Biobots, in the eyes of their proponents, could serve as sentinels, guardians, and saviors of the environment.

Researchers also vaunt the superior technical performance of biobots. These living machines hold the promise of vastly outperforming their artificial counterparts because they combine the capacities of living beings with real-time control by humans. In many instances, animals and insects currently outperform their robotic counterparts in seemingly simple tasks—such as terrestrial locomotion—by orders of magnitude. This is unsurprising, as living organisms have had millions of years to evolve and optimize the capacity for fine movement and energy efficiency. Recognizing this fact, scientific interest has grown in harnessing, rather than merely attempting to mimic, their innate capacities.

In contrast to artificial, mechanical robots, biobots have low energy consumption and low hardware costs. By harnessing the locomotion patterns of living creatures, biobots can navigate more easily than mechanical robots through terrestrial and aquatic environments.[8] Capacities that elude artificial robots—such as precise yet fluid actuation of muscles—are innate in biobots. And the sensory capacities of living organisms—visual, audio, and tactile—still far outstrip anything humans can build. Rather than trying to imitate a moth's sense of smell through, say, creating a "digital nose," why not harness a real moth's sense of smell to a computer?

DIGITAL DOMESTICATION

Our ability to digitally domesticate life has taken a leap forward. In the past, biorobotics largely limited itself to imitating nature in digital form. The concept inspired an entire field: biomimicry, learning from and then imitating nature to improve on human creations. The living world acted as a source of inspiration for building better mechanical robots. Insect eyes inspired better artificial vision,[9] while the behavior of insect colonies inspired social robots and swarm robots, which can communicate, coordinate tasks, and self-organize synchronized behavior. Some quadrupedal robots, such as

Boston Dynamic's ferocious-looking BigDog, imitate four-legged animals' gait, allowing them to carry hundreds of pounds with stability and, unlike two-legged robots, spring easily in any direction. There are also hexapedal robots that can scurry just like real insects; for fast movement relative to body size, six legs are far better than two or four. One of the first biomimetic robots, RoboLobster, was designed to scuttle and scrape the seafloor just like its biotic counterparts, to clean up underwater oil spills. Biomimicry has also inspired human innovation in "green materials," such as synthetic leaves, super-adhesives (mimicking geckos' notorious "stickiness"), and self-healing surfaces.

But scientists are now moving beyond mere imitation. A new generation of biobots (also referred to as cyborgs, animal-robots, living robots, and biorobots) merge natural and artificial worlds, both physically and cognitively. Nonhuman species that were previously undomesticated can now be controlled and directed in accordance with human desires. Some argue that this is beneficial, but others argue that this is unethical exploitation.[10]

When we tame a wild animal, we develop a relationship with it; domesticated animals may live in our homes or inside buildings that function like homes. Domestication, in other words, implies a sense of community. The roots of the word "domesticate" embody this sentiment: "dom" comes from the Latin *domum*, or "home." When we domesticate animals, we invite them into an extended human community—albeit a community infused with power relations.

Biobots don't require this same type of relationship. At the level of the individual organism, biobot engineering entails fusing engineered components with living tissue, by incorporating either an artificial component into an animal or a biological organ into a robot. This enables direct interaction between the nervous or muscular system, electronics (e.g., to control movement), and computational systems (that might harvest data or enable hybrid forms of cognition). Researchers can tell cockroaches where to scuttle, command geckos to jump, control how moths fly, and choose where fish swim. But no relationship-building is required, no interplay between human and animal that domestication entails; control is computational, abstract, and, in some cases, autonomous.

Researchers have, for example, developed fine-tuned systems for controlling cockroaches by applying electrical stimuli to their antennae or to their leg muscles (via miniature electronic components fused to or implanted in their bodies), with feedback via a brain-computer interface. Equipped with tiny accelerometers and gyrometers, the cockroach cyborgs' speed, step length, walking gait, and direction can be controlled and tracked manually via trackball, by software programs that respond in real time as the insects move around, or even by a direct brain-to-brain interface between human and cockroach. In the brain-to-brain interface experiment, both humans and cockroaches were hooked up to a shared computer. The human experimenters were able to successfully elicit a reaction from cockroaches over 80 percent of the time, although the cyborg could be steered successfully along preset tracks only 20 percent of the time (the cockroaches "heard" what people were thinking but didn't necessarily cooperate).[11]

PLANT BIOBOTS

Animals were the first frontier of biobot research, but scientists have more recently turned to plants as a source of biobot raw material. Researchers have, for example, created nanobionic plants with nanosensors in their cells.[12] These nanobionic plants can be engineered to develop new capacities, like emitting light in case of a water shortage. In one experiment, researchers at MIT engineered the nanosensors to produce a near-infrared fluorescence that can be imaged using a small infrared camera connected to a Raspberry Pi. When the plants respond to stresses such as infection or injury, their natural signaling mechanisms are detected, and an alert is sent to a smartphone.[13] Today, we have baby monitors in our houses; tomorrow, we will have geranium, orchid, and fern monitors.

In fact, there are several reasons to think that plant biobots are likely to proliferate faster than their animal counterparts. Perhaps most obviously, they are easily and widely available. The dominant life form on the planet, plants account for over 80 percent of the world's living biomass. In contrast, humans account for less than a hundredth of a percent of the world's living biomass. Moreover, the same nanosensors can be used in a wide range of plant species. Plant biologists have traditionally focused molecular biology research on plants amenable to genetic modification, which required detailed and sometimes time-consuming analyses and then manipulation of the specific genome of each species. In contrast, generic nanosensors can be injected or inserted in nearly any plant (and, thus far, have largely evaded the regulatory and public scrutiny to which genetic modification of plant species has been subjected). By combining plant organelles and nonbiological nanostructures, researchers can boost plant biology at scale. The applications for environmental purposes are ubiquitous and easily scalable because of plants' relative hardiness and low cost.

Plant nano-biobots could, for example, be used as environmental sensors. Using carbon nanotubes, the MIT researchers modified plants to detect nitric oxide gas by emitting near-infrared fluorescent light in real time, thereby acting as a photonic chemical sensor. "Nanotechnology," reads the tagline from MIT's publicity release, "could turn shrubbery into sensors for explosives."[14] Beyond being discreet, such environmental sensing devices would have an additional advantage: they would be biodegradable and nondestructive to the environment, in contrast to manufactured digital sensors, which increasingly litter our landscapes. These systems would also enable continuous, real-time monitoring, unlike the patchy monitoring systems on which we currently rely. Surveillance applications are reportedly being developed for military purposes: DARPA's Advanced Plant Technologies research aims to use plants to track soldiers deployed on foot—raising questions about the deployment of these types of technologies in illicit surveillance of civilians.

WHEN SILICON MEETS CARBON

What if plants could replace components in our digital devices and machines—whether fuel cells, batteries, or even cameras? While this research is still in its infancy, scientists have identified several promising innovations that point in this direction. Plant biobots can self-repair, grow autonomously, and generate their own energy—a perfect combination of capacities for in situ environmental applications of all kinds. Plant-inspired hybrid materials, which combine plant organelles and new materials, have been shown to exhibit mechanical and electrical properties that can rival or even surpass those offered by existing electronics. For example, chloroplasts function with high efficiency in biofuel cells. They can also be used to create self-repairing hydrogels.[15] This creates the potential for their use in biofuel cells and in bioelectronic circuits. Plant-based hydrogels might, for instance, be used as a substitute for graphene or polymer-based hydrogels used in fuel cells, supercapacitors, or lithium batteries.

Plants are also a highly attractive digital technology platform. They are prolific, adapted to nearly every environment on Earth, and self-repairing. They grow autonomously. They provide their own water distribution and power systems. They are environmentally stable outdoors, even in the harshest of environments. They have a negative carbon footprint. To researchers who believe that we need to use technology and materials that have much lower embodied energy and generate much less e-waste than the devices we use today, plants seem like perfect candidates.

By enhancing plants with nanotechnology and connecting them to digital technologies, spatiotemporal information collected from plants can be circulated via the internet, enabling an unprecedented degree of information flow between humans and plants. For example, agricultural management could be enhanced by interfacing plants with functional nanomaterials, allowing detailed, real-time information about crop health to be sent to farmers' smartphones. Wearables for plants that detect humidity and glucose already exist; combining these with nano-enhanced plants and Internet of Things applications is likely within the next decade, a development that may help solve pressing food security challenges.

Researchers hope that, one day, plant biobots will not only monitor environmental pollution, pesticides, fungal infections, and exposure to bacterial toxins, but also provide a novel energy-harvesting solution to help with the transition beyond fossil fuels. Plants will be widespread in the landscape—perhaps ubiquitous on the walls and roofs of buildings—as self-powered environmental sensors and energy-harvesting apparatuses. Nanobionic plants may even be used as sustainable alternatives to, and eventually replace, existing batteries and everyday electronic devices.

Many questions remain, however. What is the long-term viability of plants after integration with bioelectronics and nanomaterials? How would these modifications affect the complex and nuanced communication mechanisms that currently exist within and between plants? And the biotoxicity, distribution, and fate of nanoparticles within plants need to be better understood: What would happen if you ate a nanoparticle-enhanced salad?

MORE BIO THAN BOT

Today's biobots are mostly composed of living creatures, augmented with digital components. But the converse is also occurring. Researchers are bringing artificial robots to life—literally—by incorporating targeted types of living tissue into mechanical robots to enhance their performance. The field of biohybrid robotics is entering a revolution in both design and materials, drawing on biochemical engineering, genetic engineering, developmental biology, and a new generation of biocompatible electronics. Scientists are now able to exploit the functions of living cells in artificial robots. Will this be a pathway for development of biohybrid robots in the future?

The answer to this question will shape the future of robotics. Many roboticists believe that the most innovative robots of the future will be "soft robots," unlike the rigid, metallic robots of the twentieth century.[16] Instead of C-3PO, think Baymax. Like the cuddly animated film star, soft robots will be able to squeeze, stretch, climb, grow, and morph, alternately stiffening and softening.

To build soft robots, scientists combine robotics technology with tissue engineering, biomaterials, miniaturized electronic controllers, and sensors. All of these have expanded the roboticists' design toolkit, which now includes stretchable electronics and multifunctional materials for robot skins.[17] The mechanical components and electronics in these robots will be created from fluids, elastomers, gels, and other soft materials (much like the human body). These bio-inspired innovations push the boundaries of both the shapes and abilities of robots.[18] In the future, such robots may be able to perform delicate tasks that robots are currently unable to perform safely. Living materials will push the robotic frontier even further.

The problem of robot movement is also being at least partially solved by turning to the natural world for inspiration. The ability to create biomorphic soft robots depends in part on the ability to create fluid movement in a mechanical device, and here soft robot inventors face a barrier. Considerable progress has been made in the past few decades in developing movement in robots, but the fluid movements that living beings carry out so gracefully continue to elude robot designers. Smooth, precise movements over short distances are particularly challenging: a toddler can easily pick up a grain of rice with two fingers, but most humanoid robots would find it challenging or impossible to do so. Roboticists refer to this as a problem of "actuation" (the act of setting a device in motion). Although at the macro scale mechanical systems outperform living creatures (pneumatic and hydraulic systems in terms of stress, dielectric elastomers in terms of stroke, and piezoelectric motors in terms of power density and efficiency), living cells outperform mechanical systems at microscopic scales.

Why? In living creatures, movement is enabled by a biological analog of microscopic motors. At very small scales, biologically powered movement is generally superior to mechanical motion in terms of precision and controllability. Many living organisms can apply very smooth movements over very small spatial scales. Robots can't do these types of things easily. But by interfacing living cells and tissues with artificial components, roboticists are attempting to overcome these challenges.[19]

Living cells have other advantages: cells are relatively energy-efficient, self-repairing, and biodegradable. Most mechanical actuators can't be scaled down to microscopic size without losing fine-grained controllability, force, and torque. Evolution accomplished that design task for living cells millions of years ago.[20]

Take, for example, a muscle cell. In contrast to a robot, a muscle cell uses a range of inexpensive and readily available fuel sources (such as sugars and fatty acids), is inaudible when operating, generates biodegradable and hence environmentally friendly substances as it converts fuel to work, has naturally embedded sensors (muscle spindles), and can easily stiffen or relax. And all of this at a microscopic scale: several orders of magnitude smaller than the smallest piezoelectric motors. For those designing microrobots/

nanobots with onboard propulsion (as in targeted drug therapy), the ability to efficiently contract (and hence produce force) at very small scales is a necessity. So far, living tissue still outperforms artificial components. Moreover, "green materials" offer distinct advantages over artificial ones, including lower embodied energy in production, and less (or zero) e-waste. For this reason, many of the soft robots of the future are likely to be more living than artificial, more bio than bot.

THE TINY ROBOT WILL SEE YOU NOW

As the wave of biobot innovation has expanded, microscopic biobots have become a site of intense innovation. The vision of the 1960s scientific thriller *Fantastic Voyage*—in which a miniaturized submarine and its crew are injected into the bloodstream of an injured scientist to repair damage to his brain—is now a reality. A number of tiny robots have been developed to dive into the human body and deliver targeted drug therapies or combat infection. One group of robots is entirely artificial. These robots are sometimes soft, like the hydrogel-based "squishy clockwork" robot invented at Columbia University.[21] Others are more conventional mechanical devices, like the tiny "jackhammer" millirobot designed to penetrate brain tissue.

These artificial devices are stymied by serious technical challenges, however, such as scaling down below a certain size while maintaining controllability, movement, and the sensitive detection required to deliver precision doses of drugs. Researchers have tried various combinations of living cells and mechanical devices to produce micro-biobots, like attaching sperm cells from bulls to magnetic microtubes.[22] These artificially enhanced sperm biobots can deliver microscopic payloads of medicine or monitoring devices.

The desire to combine strength, suppleness, agility, and propulsion at microscopic scales has even sparked an interest in using bacteria to develop biobots. The master swimmers of the microscopic world, bacteria outperform nearly all other motile cells. Appropriately engineered, bacteria can be injected into the bloodstream or other fluid-containing areas within the body to perform therapy. Once researchers master how to control bacterial biobots in the body (a nontrivial challenge), their use in medicine is likely to become widespread. These micro- and nanoscale robots can move freely within the bodies of other living creatures. They can communicate with one another and respond to signals to perform specific tasks. Then, when their mission is accomplished, they can biodegrade. Although drug delivery is the most promising application to date, other applications in the pipeline include precision surgery, sensing of biological targets, detoxification, and tackling of antibiotic-resistant infections.

For now, these biohybrid robots are unlikely to scale up. Living cells don't exist in isolation. Without fibroblasts and vascular cells, muscles don't perform well (or even live) for very long. Living bodies are complex communities within which single cells are designed to thrive. Scientists cannot replicate such complex communities with mammalian cells. Currently, cells harvested from live animals work better than lab-cultured cells, but this raises ethical issues: Should cells from living animals be used to create biobots?

Using insect cells may partially resolve some of these issues. Insect cells can live for much longer—up to several months. Insects are tough, and far better than mammals at surviving in extremes of heat, cold, high salinity, or low oxygen. Insect biobots could conceivably thrive in almost any environment.

When in flight, insect muscles produce the highest sustained power output per mass of any animal. And although this is patently unfair, fewer people are likely to feel troubled about harvesting cells from a cockroach than from a cow. The names of insect biobots—like the "SmelliCopter," created by fusing a tiny nano-drone with an antenna that has been surgically removed from a hawk moth—often suggest a blithe indifference to the fate of the living organisms that have been sacrificed.[23]

The scholarly research literature on insect biobots also features a worrisome blind spot: it rarely mentions the fact that insects are disappearing at an alarming, indeed unprecedented, rate from the surface of Earth. One recent review indicated that significant declines were occurring in 40 percent of insect species globally.[24] As Harvard biologist E. O. Wilson once noted: "If all mankind were to disappear, the world would regenerate back to the rich state of equilibrium that existed 10,000 years ago. If insects were to vanish, the environment would collapse into chaos." Given the myriad roles that insects play in our ecosystems, from pollination to soil enhancement to nutrient cycling, it seems tone-deaf, if not pathological, to begin treating them as yet another resource to manipulate and exploit.

STRANGE ROBOTS

The original robots, first named in a play written by Czech playwright Karel Čapek in 1921, were made not from circuits and electronics but rather from flesh and blood. Čapek's robots were not made but born, distilled from flesh-like dough and tissue cultures. Much like Victor Frankenstein's monster, assembled from an alchemy between secret chemicals and discarded body parts, the robots inhabit an uncanny in-between world: neither dead nor alive, neither human nor machine. One hundred years later, Čapek's vision was brought to life: a team of researchers created a computer-designed organism (CDO), devised by an artificial intelligence algorithm, and built from the DNA of the African clawed frog *Xenopus laevis*.[25] They named their invention a xenobot, from the Greek *xenos*, meaning "stranger," "foreigner."

To build xenobots, the scientists asked a deceptively simple question: Could an AI algorithm design viable life forms? With funding from the US military's DARPA Lifelong Learning Machines program and the National Science Foundation, the researchers fed the algorithm specific constraints, including the biophysics of frog skin and heart cells (such as the maximum muscle power of their tissues) and the desired tasks to be performed (such as locomotion). To mimic evolution in the natural world, the algorithm was instructed to delete the least successful forms, while the best-performing simulated bots "reproduced" inside the algorithm. Over several months, the AI algorithm tested several thousand designs to determine the best candidates for testing with real cells.

The computer scientists then handed the most promising designs to a team of biologists who brought them to life using microsurgical techniques, wrapping frog skin cells around beating heart muscle cells, and then (using tiny forceps and even tinier electrodes) carving out the structure dictated by the algorithm. Assembled into a shape not dictated by their DNA, the cells nonetheless began to work together. The contractions of the heart muscle cells, contained in the supporting architecture of the skin cells, enabled the xenobot, composed of about 5,000 cells, to move forward. Videos of the tiny organism—less than one millimeter in diameter—waddling across the microscope viewfinder went viral.

Xenobots, preloaded with food (fats and proteins), can survive for weeks—and longer with more nutrients. When the scientists sliced some nearly in half as a test, they were able to self-repair, "zippering" their cells back up. Placed together with their kin, the xenobots would move around in circles, spontaneously and collectively pushing pellets into a central location. Other versions were built with pouches that could successfully carry an object. Unlike most biobots, frog xenobots do not require external stimulation from a remote control or bioelectricity. Once assembled, they do what they are designed to do: walk like a wind-up toy. The xenobots also demonstrated emergent properties, such as self-replication—a capacity neither designed nor anticipated by their inventors.

This autonomy creates a novel way for such computer-designed yet entirely biological organisms to be used for medical and environmental applications, like intelligent drug delivery, scraping out of plaque in arteries, or gathering of microplastic in the oceans.

What are we to make of such a creature? Scientists have known for decades that certain embryonic skin cells—called the "animal cap"—will grow cilia under the right circumstances. Prior studies had already demonstrated that *Xenopus* animal caps could, if manipulated in specific ways, grow into other tissue types—neurons or muscle. But the AI-designed xenobot did something unexpected. The designers of the xenobot argue that these entities are autonomous organisms: they can self-heal when damaged, signal to one another via pulses of calcium ions, and organize individual cells into piles that, after a few days, self-assemble into other xenobots. This latter capacity does not constitute reproduction, as no genetic material is involved; instead, the researchers termed this "kinematic self-replication."

Humans have been transforming organisms since at least the dawn of agriculture. Cloning and gene editing are now well established—if controversial—technologies. Much debate has been devoted to the ethics of applying these technologies to humans and animals. But this xenobot is different. It is neither a known species of animal nor an artificial organism. It has 100 percent frog DNA, but it is not a frog. It is the product of AI-driven evolution, not natural selection. The xenobots represent a new, different body plan, a different developmental plan—almost as if the tape of evolution had been replayed, but had come up with a new result.

As the study's authors put it, such a xenobot falls in an entirely new class of being: a living, programmed, and reconfigurable entity. But at what point does engineered tissue become an autonomous organism? Lacking heredity as an aspect of reproductive capacity, which might be deemed a requisite feature of organisms, xenobots fall into a category of their own: living, multicellular, adaptable enough to support viable life forms that are totally different from the ones that their genomes evolved to create. As the authors later noted, the classification of such a being is so controversial that even they themselves do not agree on which terms should be used: Living machine? Biological robot? Computer-designed organism? Biomachine? Self-motile biological machine?

In the future, entities akin to xenobots will proliferate. The methods are accessible, and the authors have posted the instructions under a Creative Commons license on GitHub, an open invitation to biohackers and self-proclaimed do-it-yourself biology hobbyists.[26] This emerging science has a name: synthetic morphology. Biological robots will be constructed from a range of counterintuitive components. To name just a few that already exist: pinecone and oat seed robots, engineered jellyfish, and liquid droplet robots. Cardiac-powered biohybrid designs call forth capacities from living tissues, fused to soft scaffolds, to create autonomous beings designed by AI and built—but not completely controlled—by humans.[27] The pantheon of chimeric and synthetic organisms is growing rapidly, and in a world where biopunk meets cyberpunk meets neuropunk, scientists' inventions are stranger than science fiction.

FRANKENSTEIN, TALOS, GOLEM

In robotics circles, the term "uncanny valley" has circulated since the 1970s. It refers to a curious fact about our relationship with robots: the more humanoid that robots or computer interfaces become, the eerier they seem. The new generation of ultra-realistic androids, animated characters, and avatars seem both lifelike and "not quite right." The uncanny valley is the place where that disquiet emerges: an in-between world between life and not-life.

Roboticist Masahiro Mori, who coined the term, first described the phenomenon to explain something surprising: why people reacted more negatively to his robots the more sophisticated they became. As his robots became more and more realistic, he observed, something strange happened.[28] He, like other inventors, expected people to feel a greater sense of affinity for the prosthetic hands and humanoid robots as they became increasingly lifelike—but the opposite happened. A realistic prosthetic hand often provoked a feeling of distaste and unease, perhaps because of the combination of a lifelike hand and a cold, clammy grasp. This feeling has only deepened with the new generation of ultra-realistic androids, like Hanson Robotics' Sophia or Soul Machines' Ava. Computer animations sometimes evoke similar feelings. The 2019 versions of *The Lion King* and *Cats*—which featured lifelike computer animations of talking lions and singing cat-humans—were simultaneously mesmerizing and creepy.

The concept of the uncanny has a long history in psychology, where it describes the experience of observing something apparently new yet oddly familiar. Freud observed this strangeness in the mundane when he wrote about the eeriness of dolls and waxworks. Encountering the uncanny creates unease and anxiety. A corpse might provoke such a feeling—lifelike yet not living. The feelings of fear and dread that such living/nonliving entities provoke explains why they feature so often in horror movies: killer humanoid machines in the *Terminator* movies, or replicants in *Blade Runner*. In psychoanalysis, the uncanny is omnipresent, part of the human condition. In robotics, however, the uncanny valley is a temporary state to be vanquished by innovation. Masahiro Mori depicted the uncanny valley as a brief dip in the innovation curve. His graph—which has become a meme in the robotics community—implies that robotics will eventually climb out of the uncanny valley and, one day, provoke a strong feeling of affinity in real humans.

Yet concerns about the power of lifelike robots go far beyond mere feelings of unease. With the onset of the Industrial Revolution, tales about the power and perils of fusing biological life with technology proliferated, from Mary Shelley's *Frankenstein* to the wicked robot Maria in the silent film *Metropolis* (1927). In fact, myths about biobots have their roots in antiquity.[29] Android warriors first appear in Indian historical texts in the fourth century BCE as fierce soldiers (*bhuta vahana yanta*, "spirit movement machines"), wielding whirling blades, serving kings, and guarding Buddha's tomb. At about the same time, tales of androids also appear in China: man-made men who talk, walk, dance, sing, and perfectly mimic the actions of real human beings. The bronze android Talos—a self-moving automaton crafted by Hephaestus, god of the forge and technology—guarded the island of Crete, marching around the perimeter three times per day, hurling stones at approaching boats, and crushing strangers who arrived on the shore with a deadly embrace. These myths have counterparts in contemporary science fiction. In the latest iteration of the *Day of the Triffids* (a TV series based on the 1960s bestselling novel), the triffids are a naturally occurring species from Zaire that have been selectively modified by scientists as an alternative to fossil fuels, to avert global warming. Inadvertently, the scientists create a hybrid that attacks rather than saves humanity (in an earlier Cold War adaptation, the man-eating triffids were created by Russians). The familiar trope of an android being brought to life only to overthrow its masters—*Blade Runner*, *Ex Machina*—takes on a biological twist.

Myths express our fears, even terror, of what nature-machine hybrids might do to humanity. But only rarely do these myths ask about the effect of biohybrid life on nonhumans. This is a peculiar gap, mirrored in contemporary ethics. As a field of research, bioethics has largely focused on technological advances in medicine and biology as applied to human life. Contemporary bioethics, from this perspective, provides largely legalistic guidelines for health policy but does not question the implicit species hierarchy at the core of modern science and medicine. Hence, although bioethicists debate hybrids made with human cells (chimeras), they have not devoted as much attention to hybrids between nonhuman species and machines.

Environmental ethics has a similar gap. Ethicists concerned with the environment tend to restrict their focus to non–technologically manipulated organisms and landscapes. Interspecies biohybrids that do not include humans tend to fall between the cracks. When human genetic material is mixed with that of other species, this is referred to by ethicists as "human-animal chimeras." Ironically, regulations delineating new boundaries between humans and other animals have proliferated, despite growing scientific consensus that conscious life is a continuum (and despite arguments from some quarters that challenge the sharp dividing line that we draw between species). Indeed, a recent spate of new guidelines in several countries draws an even sharper line than in the past between human genetic materials and those of other living creatures.[30]

In seeking to govern chimeras, regulators are now confronting the uncanny valley, and their stance is deeply contradictory. On the one hand, they continue to delineate the distinction between humans and other animals; debates and decision-making invoke concepts like the sanctity of the human. On the other hand, they permit and endorse the creation of what they often term "human-animal chimeras"—which call into question the divide between humans and nonhumans.

In taking this approach, regulators and bioethicists reveal a deep-seated commitment to speciesism. Much of the current debate on human-animal chimeras focuses narrowly on a subset of experiments using human stem cells.[31] Tighter regulations have been applied to the ethics of incorporating animal cells into humans, but rather less attention has been paid to incorporating human cells into animals (e.g., engineering nonhumans to grow human organs to be harvested for transplants). In contrast, animal-animal and animal-machine chimeras have been less tightly regulated. Nonhuman interspecies mixing does not yet inspire the same degree of scrutiny and debate as human-animal chimeras. As Michael Levin, one of the designers of the xenobot, puts it:

> Human society has a long history of moral failures toward numerous types of human and nonhuman animals based on distinctions that are now widely recognized to be irrelevant. By forcing us to search for deeper essential meanings of categories that drive moral concern and responsibility (agency, intelligence, etc.) in unfamiliar novel guises, we will be better placed to more appropriately implement compassion in our relationship to a truly diverse community of beings.[32]

How far does our circle of care extend when considering chimeras? The word we use to frame this question, "chimera," already partially frames our answers to these questions. In Greek mythology, Chimera was a monstrous creature composed of the parts of several animals. Often depicted as a lion with the head of a goat protruding from its back, and a tail ending with a snake's head, Chimera could breathe fire like a dragon and was the sibling of the Greek pantheon's most feared monsters: Cerberus (the hound of Hades) and Hydra (the many-headed serpent monster whose breath was poisonous, and which would regrow two heads if one was chopped off). The siblings' heritage expresses fears of species intermixing, as they are descendants of Echidna (half beautiful maiden and half fearsome snake); the serpentine giant, Typhon; the earth goddess, Gaia; and the ancient netherworld king, Tartarus.

Arguably, each generation has its own unique frontier of technological anxiety. Yet the fluidity of this frontier does not reduce the emotional charge it holds. And our conceptual and regulatory frameworks seem ill-equipped to handle this proliferation of species mixing in our pursuit of artificial and augmented life. Our regulatory frameworks are also silent on the potential for transformative and perhaps irreparable damage to the genome of Earth's living species and the ecosystems on which they depend. And they pay scant attention to new forms of discrimination. Biohybrid robots and organisms open the door to a new eco-industrial paradigm that, according to its advocates, offers a greener future, moving beyond the domination of nature through extraction of resources and generation of pollution by artificial machines. This vision positions biomimicry and bioenhancement as utopian technologies that leave behind the crude violence of the Industrial Revolution. As the xenobot designers argue, their work allows scientists to go beyond the contingent natural products of evolution. Humanity now possesses the technology—as biologist Stephen Jay Gould once said— to replay the tape of life, and explore the radically different evolutionary outcomes that may result.[33]

But what if, instead of useful innovations, we are creating a new generation of biodigital Frankenstein's monsters? Biohacking, from this perspective, sits on a continuum with twentieth-century crop science, which sought to deliver benefits (like crop yield increases) through the Green Revolution, but inadvertently compromised water quality and insect health, feeding humans at the expense of biodiversity. Is biohacking a tool for environmental improvement, a weapon of environmental destruction, or both? Is it admirably creative, or rather a violation of dignity, to create biobots? A useful form of green technology or a form of torture?

SYMBIOGENESIS 2.0

Projected into the future, the logical conclusion of the collective of biodigital technologies is a seeming contradiction: as digital innovation penetrates the lifeworld, it becomes increasingly biological, and digital technology as we know it will cease to exist. As our biosphere becomes computational, living organisms merge with the digital tools created by *Homo sapiens*. This insight builds on the work of biologist Lynn Margulis. Working with NASA in the 1960s on the search for life on other planets, Margulis became intrigued by the history of life on Earth, and in particular by one of the most profound questions in the history of biology: How did ancient life make the leap from single-celled to multicellular organisms? At the age of twenty-nine, she published her answer: the theory of symbiogenesis. Margulis's theory holds that the evolution of complex, eukaryotic cells (which have a nucleus and complex organelles) occurred through the fusion of different bacterial and archaeal cells. Sometime in the distant past, simple single-celled organisms merged together in symbiotic relationships, leading to the development of a new, more complex organism. Margulis proved her theory through investigating the genetic makeup of cell organelles such as mitochondria (in animal cells) and chloroplasts (in plant cells), which, she found, have their own DNA and have structural similarities to bacteria. Published in 1967 after fifteen rejections, her germinal paper provoked controversy, but her ideas eventually became widely accepted.[34] Symbiogenesis—the coming together of different organisms to create new, viable ones—remains one of the most compelling ideas in the history of Western science.

Margulis's work revealed that our cells and bodies are composed of elements that were once multiple species. Long ago, our mitochondria were once separate organisms, and their distinct DNA, different from our own and inherited solely through the maternal line, has been preserved within us.[35] And, as Margulis observed, this meant that collaboration as well as competition shaped the evolution of life on Earth.

Researchers have continued to explore this concept of symbiosis. Recently, Margaret McFall-Ngai and twenty-four coauthors presented a systematic review of animal-bacterial interactions from cellular to ecosystem scales, in which they argue that bacteria and animals are symbionts: bacteria have facilitated the origin and evolution of animals, and influenced animal genomes and animal development. Biologist Scott Gilbert argues that this symbiotic view of life means that "we have never been individuals." We are biological collectives: from cells to our whole bodies, from ecosystems to the whole planet.[36]

Margulis anticipated these ideas in the 1960s, when she began developing the theory of Gaia along with James Lovelock. Earth's biogeochemical processes, argued Margulis, were akin to the symbiotic processes she observed in the tiniest cells. Gaia theory holds that living and nonliving components interact and are interconnected by cycles of nutrients, water, and energy. To some extent, these cycles are self-regulating; living organisms thereby influence, and perhaps even regulate, the planet's atmosphere and temperature. As one of her students quipped, Gaia was just "symbiosis viewed from space." Within our cells, as well as across the planet, different organisms with different DNA interact in mutually beneficial ways. Living things contribute to nonliving things that shelter and protect them—like the atmosphere. Our atmosphere owes its existence to cyanobacteria (blue-green algae), which first began to oxygenate the atmosphere over 2 billion years ago, creating the conditions for life as we know it.[37] Because Earth's atmosphere was almost entirely made by living things, Lovelock once likened it to the fur on a mink, or the shell on a snail. Symbiotic relationships between different aspects of our ecosystem influence planetary conditions, including those necessary for life as we know it: the atmosphere, the cycling of nutrients, the movements of tectonic plates.

Margulis's theory of symbiogenesis also suggests that symbiotic relationships between organisms can lead to the emergence of new, more complex life forms. From this perspective, the biodigital revolution is another chapter in the unfolding book of symbiogenesis. Our bacterial ancestors absorbed other bacteria, and learned to collaborate in a mutually beneficial way. This gave rise to complex life forms, including humans who invented digital tools. Having invented these tools, humans began to use them to manipulate the code of life itself. Our current era—christened the Anthropocene by scientists—is one in which humans have become a geological force at a global scale, influencing the planetary climate. This is due to innovations that are centuries old, an outcome of the Industrial Revolution.

Innovations like the xenobot are harbingers of a new era, in which humans could become a biodigital force at a global scale, reshaping life on Earth. Biodigital innovation enables the fusion of biological and digital technology, which has the potential to create new forms of symbiotic relationships between living organisms and digital devices. Synthetic biology entails the design and construction of organisms and biological systems; those who master both digital code and genetic code become the new creature-makers. The fusion of humans, nonhumans, and machines is no longer the stuff of science fiction. Humanity is now poised to become, for better or worse, a tool-*infused* species in addition to a merely a tool-making (and a tool-wasting) one. As scientists create animal-machine hybrids and entirely new life forms, life begins to ingest and merge with its tools: Symbiogenesis 2.0.

This convergence of the biological and the digital is occurring at the level of individual organisms but also, as explored in the following chapter, at the level of buildings, supply chains, and ecosystems. As the distinctions between environment and organism, human and animal, living and machine become increasingly blurred, conventional questions of environmental governance become less relevant, and perhaps less tenable. At the moment of apparent digital supremacy, the disappearance of the digital as we know it is on the horizon. Questions of environmental governance become increasingly digital; and questions of digital transformation are inescapably biological and ecological.

PARABLE OF THE ENDLESS FORMS

Once we counted on our fingers
and tried to enumerate the stars
notches and pebbles.
Once we calculated on our machines,
ledgers and vectors.
Once we conjured our chimeras,
Manticore and Sphinx.

Now our cells compose equations,
proteins and genes.
On a planet that computes energy,
sunlight, and air.
We convert signals and symbols
into biobots and cyborgs.

If computation is embodied
If all matter is malleable
Which forms shall we assume?

10 WHEN ALL THE COMPUTERS MELT INTO AIR

At the turn of the twenty-first century, scientists have endowed humanity with a new capacity: tools to manipulate DNA, the code of life. To understand the significance of this development for environmental governance, let's step back and briefly consider the arc of human history, in which two major innovations profoundly expanded our ability to transform our planet and metabolize its resources.

The first innovation dates back to the Neolithic period, an era of domestication when humans learned how to tame fire, animals, and plants. The second, more recent innovation is the Industrial Revolution, when humans learned how to accelerate mechanization through fossil fuels and invented technologies that harnessed planetary processes—such as nitrogen fixing—at an industrial scale. At each step, technological innovation and the ability to control energy and other organisms accelerated human colonization of Earth. Domestication and mechanization enabled humanity to colonize the planet and planted the seeds for the Great Acceleration—the rapid increase in human population and metabolism of Earth's resources over the past century.

Today, we are on the cusp of a biodigital revolution, in which scientists use digital technology to read, write, and execute biology. As explored in this chapter, synthetic biology—which involves redesigning ecosystems and organisms through manipulating DNA—is being used by scientists to create novel forms of life and human-engineered ecological interactions. Advances in gene editing, artificial intelligence, and materials technology, from living and smart materials to biocompatible electronics, promise startling new powers: the ability to harness energy from photosynthesis, to rewrite the genetic code of entire species, and even to embed computation into living organisms. Yesterday, the mastery of fire; tomorrow, the mastery of evolution.

In the biodigital Anthropocene, the ambivalence of humanity's toolmaking abilities is acute. These innovations may enable greater exploitation of life or, if used with caution and care, may renew our relationships and foster new collaborations with other species—which one day might build our dwellings, own property, and perhaps even vote. I refer to this latter view as "digital biocentrism," a worldview where digital innovation imitates and embeds itself in biological and ecological processes. It may seem counterintuitive, in an era when digital networks are spreading throughout the globe, to predict the end of purely digital innovation. But as biodigital innovation is used to manipulate cells, bodies, and ecosystems, it requires the insertion of digital tools into living beings, blurring the distinction between humans and nonhumans and between living organisms and machines—to the point where digital technologies, as we recognize them today, may one day disappear.

THE BIODIGITAL REVOLUTION

What, exactly, is biodigital innovation? Simply put, biodigital innovation entails the interactive combination (and, at the limit, the merging) of digital and biological technologies. This encompasses the use of biodigital tools—such as CRISPR/Cas9—to manipulate living organisms and their components, down to the level of individual cells and even DNA, which enables scientists to create new types of organic compounds and synthetic organisms. One aspect of biodigital innovation focuses on replicating or tweaking living systems, through the use of synthetic biology; for example, researchers have recently made significant breakthroughs in programming logic computations in single mammalian cells.[1] Another dimension of biodigital innovation combines these techniques with artificial intelligence and materials science to create organism-machine hybrids, and entirely new forms of life, with completely redesigned DNA—as demonstrated, in the previous chapter, by biobots.[2]

Biodigital innovation has three sets of implications relevant to environmental governance. First, biodigital innovation implies the coevolution of biological and digital technologies. Using digital technologies to study living organisms, from bacteria to humans, creates the potential for humanity to intervene and manipulate biology in a manner that was impossible even a few years ago. Recent advances in gene editing, for instance, would be impossible without digital technology and bioinformatics. As explored below, the use of synthetic biology to alter ecosystems through gene drives is one powerful example of biodigital innovation, which creates new opportunities and existential questions for environmental governance.

Second, biodigital innovation entails the biophysical integration of biotic entities and digital devices: the incorporation of digital inputs, outputs, or processing into living biological tissues or materials. Some scientists believe that many types of computational systems could be built using materials derived from living organisms; for example, researchers have recently developed the world's first transistor made from wood.[3] Biophysical integration may also occur through fusing computers and living organisms; robots with biological brains are one example, and biobots, in which digital technology is fused to the flesh and nervous systems of living organisms, are another. But biodigital integration may also occur at the systems level, from urban systems to supply chains to ecosystems. We will radically enhance our ability to sense, store, process, analyze, and share information by using living organisms as sensing devices (such as the many biohybrid robots discussed in the previous chapter). Through the development of living material interfaces, researchers also intend to introduce biodigital innovations into digital devices, homes, infrastructure, and cities. Scientists have developed new specializations (variously referred to as animal-computer interaction, human-plant interaction, moist media, biomedia, organismal engineering, digital living media, agential matter) that replace electronic components with biological materials. A simple example is a biodigital display that uses glowing bacteria or plants that change color, size, or directionality to communicate data. Biodigital innovators argue that their inventions will radically alter industrial production systems and supply chains, reducing energy use and waste. For example, as biochips become more widely used—particularly for medical and ecological applications—there will be less demand for rare earth minerals and other commodities that are necessary to build electronic chips. To take another example, a DNA data center made up of *E. coli* will require far less space and electricity than a contemporary server farm.

Third, biodigital innovation implies conceptual as well as technological convergence around key concepts—information, computation, energy—that are understood to characterize both biological and computer systems. The distinction between "hard" digital data and "wet" biological information will become increasingly irrelevant. In the next few decades, biodigital technologies will be woven into the fabric of our clothing, devices, buildings, and cities—and perhaps even our bodies. If the more futuristic visions of biodigital innovators come true, computation will migrate from machines to carbon-based life forms, and the computers and other digital devices that we use today will be gradually rendered obsolete. This is more than a technological change; it is also a psychological one. Biodigital technologies blur distinctions between humans and nonhumans, natural and unnatural, organic and machine. They thus transform the way that humans understand their own bodies and living spaces, economies, and societies. In the remainder of this chapter, I explore potential environmental implications of three types of biodigital innovation: synthetic biology, living buildings, and biological computing.

SYNTHETIC BIOLOGY IN ENVIRONMENTAL CONSERVATION

The use of synthetic biology for conservation purposes is one of the most controversial topics in contemporary environmental governance. Proponents argue that engineered genes, for instance, could offer significant protections to endangered species and that refusing to use this technology is unethical, insofar as we are refraining from using a tool that could help mitigate biodiversity loss. Skeptics, however, argue that using such new technologies "in the wild" could create unintended harms for the species that they are intended to help or, worse, spread to nontarget populations.

Three innovations are at the center of this fierce debate: gene drives, genetic manipulation of organisms to improve their resistance to climate change, and engineered enhancements to boost photosynthesis.

Gene drives, while highly controversial, have been proposed as a strategy for eliminating species such as malaria-carrying mosquito populations. Every year, more than half a million people—the majority of them children under the age of five—die from mosquito-transmitted diseases such as malaria, dengue, and Zika (a virus that produces significant birth defects if it infects pregnant women). This threat is growing as mosquitos and other insects expand their ranges due to climate change. Gene drives, which involve altering the genome of a species so that a particular suite of genes spreads throughout the population, offer a potential solution to this problem; for instance, gene drives can be used to ensure that a species produces only sterile offspring, and hence dies out. Using similar techniques, gene drives could also be used to control or eliminate invasive species—nonnative organisms that often have no natural predators, and thus wreak havoc on ecosystems into which they are introduced. Proponents argue that because such gene alterations are easy to detect through genome sequencing and are unlikely to accidentally spread through populations in which mating choice is tightly controlled (such as domesticated species and agricultural crops), they are likely to be safe. Skeptics point to the high degree of risk entailed in introducing self-replicating genes with such powerful consequences into the environment—treating the whole planet, in effect, as a natural laboratory for a technology with an unknown level of risk.[4]

Synthetic biology techniques are also being used as a strategy for climate change adaptation through engineering genes to enhance the capacity of plants and animals to cope with extreme heat or drought;[5] CRISPR-mediated genome engineering, for example, has been used to develop rice strains with greater heat tolerance.[6] Similarly, coral larvae have been genetically engineered in the lab to increase thermal tolerance, although the proposal to use such larvae in wild reef restoration remains highly controversial and has not yet been attempted.[7] Proponents argue that these innovations are urgently needed, given that many species are struggling to adapt to climate change; given the profound changes to the world's ecosystems that are occurring, genetic engineering offers the promise of enabling organisms to be more resilient to environmental stressors. But, as the Columbia University biologist Robert Pollack has argued, such technologies open the door to the return of eugenics: the selection of "good" versions of genomes and the weeding out of "bad" versions creates existential risks of both harm and exclusion that society should deem to be ethically unacceptable.[8]

Another contentious topic of debate is the use of synthetic biology to boost photosynthesis, which advocates claim could be a solution to climate change and the world's future energy needs. Across the Earth's surface, absorbed sunlight powers photosynthesis to produce 100 billion tons of plant biomass annually. From a global perspective, plants constitute the largest energy factory in the world, made up of quadrillions of chloroplasts that power this solar energy conversion. While photosynthesis is the fundamental process by which plants and photosynthetic microorganisms convert sunlight into chemical energy, the efficiency of photosynthesis is limited; only a small fraction of absorbed sunlight is converted.

What if scientists could engineer plants to harness the sun's energy in a more efficient manner, both improving agricultural productivity and potentially reducing the carbon footprint of human activities?[9] Plant researchers are attempting to do just that, combining nanotechnology with synthetic biology to harness photosynthesis within plants as an alternative renewable energy source. One group of MIT researchers, by injecting self-assembling nanoparticles into plants, were able to boost their ability to capture light energy by approximately 30 percent. The researchers designed single-walled carbon nanotubes that were passively absorbed directly into plant chloroplasts (the organelles where photosynthesis takes place). These nano-enhanced chloroplasts are better at photosynthesis because they capture a broader spectral range of light and scavenge more radical oxygen.

Beyond applications in agriculture, such innovations might have applications in energy systems. Although the image of suburban streets lit with glowing shrubs, rather than electrical street lights, might seem like science fiction, such technology is already within our grasp. Nanoparticles, for example, can be used to transform living plants into light-emitting objects. Salad lovers take note: spinach, arugula, watercress, and kale are apparently naturally high light emitters, which researchers have boosted with nanoparticles impregnated with luciferase, the enzyme responsible for bioluminescence in fireflies.[10] Instead of streetlamps made of steel, glass, and plastic, our streets may one day be lit with carbon nanotube-enhanced plantlamps. Plant-based biofuel cells are also in the works, as are plant-based microbial fuel cells. If these succeed, their designers hope that they could provide an environmentally friendly solution to some of our energy storage challenges.

LIVING BUILDINGS

The construction and operation of buildings accounts for over one-third of global energy use and nearly 40 percent of global energy-related carbon dioxide emissions.[11] For decades, architects and engineers have argued that we could reduce this impact by making green buildings. Although this term once conjured fantastical images—such as Milan's renowned Vertical Forest building, in which green plants nestle in the tiered façade, like shelf fungus on a tree—the idea of incorporating living plants into urban design has become the new normal of twenty-first-century architecture. By incorporating living plants in their design, green buildings are a welcome change from the "concrete conquers all" approach to building that characterized much of the twentieth century. Yet while such buildings have some obvious benefits—plants can, for instance, help with thermal insulation and mitigate the urban heat island effect—they are, arguably, merely superficial cladding for a business-as-usual approach to buildings. Green roofs and living walls do little to reduce construction waste (which is often the largest proportion of landfill wastes globally) or to significantly reduce the consumption of natural resources by the construction industry.

But what if our buildings themselves were made of living materials? And what if the animals, insects, or plants that compose those buildings could participate in designing and building them? Could digital technology enable nonhuman architects to work alongside humans, in biohybrid construction and co-design of our living spaces? How might this work in practice?

In the Silk Pavilion project, architect Neri Oxman developed a domed structure spun by silkworms. The dome's prefabricated scaffolding was robotically wound with a pattern of silk threads with many gaps. When released, the silkworms patched the gaps in the existing pattern, much as they would when building a cocoon. Gaps larger than the body size of the silkworms were left open as natural windows. The completed dome—made of densely matted yet breathable fibers that eerily recall the traps created by the Mirkwood Forest giant spiders in *The Hobbit*—lets in light and air while shielding occupants from the elements. Silkworms and humans have co-created a functional, biodegradable, and ethereal building.

Co-occupied Boundaries is a similar project. Asya Ilgün, an architect and computational designer, used 3D printing to create a scaffold specifically shaped to allow honeybees to construct their comb. The dense polymer filament is woven with gaps sufficiently large for honeybees to pass through, enabling them to access all sides of the scaffolding, and the coarse texture and sloping angles of the scaffold were tested to develop a surface to which the honeybee comb can easily attach. Ilgun's vision: a free-standing structure composed entirely of honeycomb, whose final design is controlled by the bees.

Designers of living buildings combine biology, computing, materials engineering, art, and architecture. Oxman describes this work as "material ecology": digital morphogenesis and a self-consciously adaptive environmental design. Structural design must adapt to the preferences and behaviors of the nonhuman collaborators. For animals (like silkworms or honeybees) that naturally deposit material, mechanical (and robotically enhanced) scaffolds can steer the organisms to build in specific ways. These techniques need to be calibrated to different organisms: silkworms respond to density, whereas honeybees respond to voids. By adapting the structure to the evolving needs of the animals, we co-create something beautiful: a machine/human/animal construction that none of us could create alone. Co-designed buildings thus represent an attempt to decenter the human in building design. But lurking behind the boosterism of biohybrid architecture is a lingering doubt: Is this an exploitative new form of domestication, rather than an example of multispecies cooperation? No one asked the silkworms if they wanted to build a dome, and they do not own the fruits of their labor.

Although beautiful, buildings co-created with animals via material deposition have limitations. They might support the weight of honeybees or silkworms, but not humans. And while stiffening methods (like impregnating the structures with resin) could strengthen them, this is unlikely to be a cost-efficient way to create a post-concrete world. In the future, these materials are more likely to be used for nonstructural roles, like thermal insulation, cladding, or façades.

One way to overcome these limitations is to employ plants as structural elements of buildings. There are, of course, age-old techniques: think of living willow coppicing (which has been used to create multistory buildings), arbor sculpting, and "grow-it-yourself" furniture. However, designers have recently begun designing much larger structures; christened "hortitecture," this agenda involves using trees to create actual walls and load-bearing structures. The House of the Future built by German architect Ferdinand Ludwig is a good example: the building's walls are composed of trees, with branches winding their way through glass, to create a tree-glass façade. Inside, a coiling ramp winds upward to the upper floors—an artificial yet natural treetop. The building gives a new meaning to the term "treehouse": more like a living cathedral than a backyard fort.

Such living buildings display some curious characteristics: their structural performance increases rather than degrades over time; they are highly resistant to corrosive environments; they support biodiversity; and they mitigate rather than increase urban heat islands. Some of these structures may even outperform and outlast their artificial counterparts; for example, living root bridges in India (which use the supple roots of *Ficus* trees) have been shown to outlast steel suspension bridges, due to their resistance to moisture and flash floods—sometimes enduring for centuries after their mechanical counterparts fail. Training plants to grow in such a Baubotanik (botanical construction) fashion is slow and painstaking, because designers must move at the pace at which plants grow. You can build your dream home, but it might take thirty years before you can add a second story. Given current housing shortages in many parts of the world, the idea of such slow building might not seem desirable today, but after human population peaks (which is likely to occur within the next hundred years), a stable or declining human population—not necessarily in need of more buildings but rather desiring better buildings—might have more patience.

The next generation of living architecture is undergoing a transformation via digital technology. The most forward-thinking experiments combine living plants with architectural biohybrids, cultivated through a combination of robotic automation and careful cultivation of species' individual growth patterns. Rather than inert materials (or materials that were once "lively," like wood, bricks, or soil), living buildings are co-constructed with living organisms that remain alive even after the building is completed. Collaborating with robots, these organisms also co-create the different structures within the buildings, such as load-bearing walls, envelopes, roofs, and energy generation systems. These biohybrid (or co-cultivated) building materials are self-assembling and, in some cases, self-repairing.[12]

Flora Robotica, for example, is a system that uses robot-plant hybrids to produce living, growing architecture. The system is self-organizing and uses robotic assemblers to support and control the living plants through scaffolding, water, and stimuli based on plants' distinct tropisms: phototropism (plants lean toward light sources), gravitropism (plants mostly grow upward), and thigmotropism (plants sense objects by touch, and grow around structures such as lattices and poles). As the building grows, structural features develop (walls, roofs, benches) and the plants eventually begin supporting the robots' weight. The scaffolds are braided and woven in complex forms that look like ethereal coral braided with 3D-printed kelp; flexible, biodegradable, and repositionable by tiny robots stationed at key nodes, the scaffolding is calibrated to growth patterns of different plants, and gradually becomes absorbed into the "living weave" of the building.[13] Human input is limited; rather, machine learning algorithms, combined with input from a comprehensive array of sensors that measure plant health and growth, enable the robots to learn the plants' preferences and reshape the scaffolding accordingly.[14]

The autonomy of nonhuman designers is the leitmotif of this speculative living architecture agenda. The plants, mediated by the robots, shape the overall design of the structure as it grows, creating structures that are more livable for the plants themselves. The engineers' innovations—like vascular morphogenesis controllers—enable responsive, real-time collaboration between robots and plants, which can autonomously shape their own growth, in response to (but not controlled by) human desires.[15]

Flora Robotica's creators are an interdisciplinary team: computer scientists, roboticists, biologists, engineers, and architects. They are using, and in some cases inventing, new disciplines (self-organizing swarm systems, biohybrid mechatronics, artificial life zoology, 3D biomaterial weaving). They imagine a future where entire cities have been built by biohybrid plant robotics. These living structures, as envisioned in futuristic scenarios, would monitor air quality (and clean and decarbonize the air), increase oxygen levels, create sanctuaries for insects and animals, and serve as a food source (fruits, herbs) for both humans and nonhumans.

Phil Ayres, a professor of biohybrid architecture in Copenhagen, and his colleagues frame this living architecture as a type of symbiosis: species that mutually assist rather than compete with one another. Symbiotic relationships have, of course, evolved over thousands of millennia. The question now is whether new types of symbiotic associations—complex biohybrid symbionts—can be successfully designed, rather than evolving naturally. We are still in the very early stages of such innovations. At their most speculative, researchers imagine these "mixed societies" being incorporated into every facet of material infrastructure used by both humans and nonhumans—from homes and offices to bridges and roads. The infrastructure of the future may be co-created and biodegradable, and even regenerative. Some of these smart devices might even be entirely biologically based. In a parallel to human-machine interfaces, researchers have implanted conductive nanowire inside plant xylem to create enhanced touch and motion sensor functionality and developed "infotropism" interfaces, which encode human behaviors as signals that allow living plants to act as interactive displays.[16] These innovations might eventually replace our digital devices. Plant biobots embedded in walls might, for example, serve as living cameras. Single-celled sensors can now monitor individual stomata (tiny pores on plant leaf surfaces that are highly sensitive to different wavelengths of light and open and close in response to light color and intensity). Stomata are, in other words, analogous to living light pixels. When scaled up to multiple pixels, this could be converted into a representation of an image: a plant-based analog to a digital camera.[17] These, along with other biosensors, might be incorporated into surfaces of green buildings. Decades of research on organic (carbon-based) technologies have produced working organic electronics. However, although they have some advantages (their "soft" nature allows for more flexible and thinner devices), they do not match their inorganic counterparts in terms of performance and stability—at least not yet.[18] If (or when) this problem is solved, plants could replace some of our digital devices, interwoven into the fabric of our working and dwelling spaces.

AN INTERNET OF LIVING THINGS

The idea that future smart devices might be biologically based subverts a major assumption about the future of our digital world. Until recently, innovators assumed that digital components would be artificial, that is, created from engineered, nonliving components. But what happens if we abandon this idea? What if so-called smart or programmable matter, built of materials that can perform information processing, was also alive?

Living matter is, of course, inherently "smart": capable of computation, gathering data about the environment, internal energy generation, object manipulation, and nuanced interaction with the environment. Even apparently simple unicellular organisms display complex behaviors. Scientists have documented, for example, the amazing abilities of slime mold to navigate through mazes and to find the most efficient distribution of highway and metro networks.[19] Fungi, meanwhile, carry nutrients and information between different species (through mycorrhizal networks in the soil), communicating precise information through long distances.[20]

What if computer scientists and engineers could harness living media interfaces for biological computing? What if smart matter was composed of organic, living electronics rather than inorganic, artificial electronics? This is less far-fetched than it sounds. "Green electronics" is a fast-developing field that uses natural compounds—like silk, chitosan (derived from shrimp), or albumen (egg white)—to create biocompatible electronics.

These compounds already have powerful medical applications (such as in targeted drug delivery); given that they are low-cost and biodegradable, they may begin to turn up in everyday applications, such as edible sensors for food packaging. In contrast to the Internet of Things, which threatens to create a tsunami of microelectronic waste, researchers envision a future of compostable smart matter and biodegradable smartphones: an Internet of Living Things.

Yet when it comes to the possibilities for truly living smart matter, this is only the beginning. Microelectronics, as discussed in the previous chapter, can be integrated into living tissue. These biohybrids integrate functional nanomaterials with living systems, bringing together materials engineering and biological science.[21] Freed from some of the constraints of conventional electronics (such as their physical limits at nanoscales), they can expand our sensing capabilities in entirely new directions.

But what if the microelectronics themselves were made of living tissue? Cells and their organelles, for example, can be used as components in building analogs to electronic circuits. Biomolecular computing (or biocomputing) uses DNA and other living materials as the building blocks for living computational machines. Biomolecules can be used as transducers; for example, antibodies (which are naturally exquisitely tuned) could be combined with nanoscale electronics to create ultra-sensitive detection methods. Enzyme-based cascades can be used to mimic Boolean logic operations.

DNA—life's information-storage mechanism—can also be digitalized.[22] This might seem like a quirky invention with few applications, but DNA storage is a likely solution to a looming problem. Somewhere close to two megabytes of data are created every second per person globally. Data storage systems (whether magnetic or optical) can't hold this amount of storage; analysts estimate that global memory demand will exceed projected silicon supply by the year 2040. And the energy costs (and climate emissions) associated with running these data centers are astronomical. As the world comes online, it is facing a serious data storage problem.

How would DNA storage solve this problem? Within DNA, data can be stored in long chains of nucleotides; the technology to read (sequence), write (synthesize and gene edit), and accurately copy DNA already exists.

DNA storage requires relatively little energy, and large amounts of data can be stored at densities that exponentially exceed that of electronic devices. The DNA of *E. coli* bacteria (commonly found in your gut), for example, has a storage density of about 10^{19} bits per cubic centimeter. At this density, the world's data storage needs for an entire year could be met by a cube of DNA that measures only one cubic meter in volume. And if this sounds far-fetched, Harvard researcher George Church has already used CRISPR DNA-editing technology to record images and even short digital movies into bacterial DNA, which were then read out with over 90 percent accuracy.[23]

Researchers at the University of Washington and Microsoft have gone even further: they have developed an automated system for storing data generated on digital devices in DNA. To the average user, these technologies will feel just like any other cloud-based storage; but instead of your vacation photos being on a server farm, they might be stored in a single-celled bacterium in a climate-controlled lab.[24] Rather than the RAM (random-access memory) in today's computers and smartphones, researchers are creating NAM: nucleic acid memory.[25]

Other forms of biocomputing are even more esoteric. The bioengineer William Ditto, at the Georgia Institute of Technology, built a simple neurocomputer (capable of basic addition, but not much more) from leech neurons (capable of basic addition, but not much more).[26] The discovery that mycelial networks (which are pervasive throughout the soil in virtually every ecosystem on Earth) engage in electrical signaling has stimulated debate over the viability of fungal computers. If scientists could master how mycelial networks respond to specific stimuli, engineers could treat them like living circuit boards.[27]

Andrew Adamatzky, the director of the Unconventional Computing Laboratory at the University of the West of England, has built prototype fungal computers in his lab; now he is using fungi to build a prototype smart building. Fungi are some of the largest, longest-lived organisms on Earth.[28] If they could be mobilized for environmental sensing, this would enable an unprecedented ability to monitor our environment at scale, in an unobtrusive and environmentally benign manner.

For now, fungal computers can't outcompete their silicon cousins on speed; they're far too slow, as it can take hours for a signal to propagate a few feet. But they could be a viable substrate for environmental sensing devices. Living fungi respond to many types of environmental stimuli simultaneously, in a remarkably self-organized manner, over large distances. Using fungal networks to monitor and report on soil and water quality, pollution, and other environmental features to which fungi are sensitive is not yet practical. But major research initiatives (such as the EU's Fungal Architectures project) are exploring this further, bringing together architects, computer scientists, biophysicists, and mycologists to explore mycelium-based technologies that could enable a fully integrated structural and computational living substrate using fungal mycelium for the purposes of growing architecture.

DISAPPEARING COMPUTERS

Many of the inventions discussed in this chapter might seem frivolous or on the fringe. But think back to the mid-1980s. In 1985, the first domain name (symbolics.com) was registered; that same year, Microsoft released Windows 1.0. Digital computing was just beginning to enter the consumer market. Apple had just released its Macintosh 128K—its first personal home computer. Today, biodigital innovation is at a similar inflection point, moving beyond large pharmaceutical and university research labs into widespread consumer and commercial use. The next generation of synthetic biology applications, aimed at the consumer market, may be as revolutionary as personal computers: all-in-one bioengineering devices that are as easy to use as a desktop printer—except that they can print genetic circuits in bacteria. Biohacking and do-it-yourself synbio labs have sprung up around the world—much to the concern of bioethicists.

These innovations—astounding in themselves—will one day be ubiquitous, embedded in our everyday surroundings. One example of this is called Touché, invented at Disney. This sensing technique (its official scientific label is "swept frequency capacitive sensing") is based on the same capacitive touch technology used in touchscreen displays, but with a key difference: it monitors signals across a broad range of electrical frequencies, making it possible to detect differences in gestures and even body parts (due to the differences in the conductivity of different body tissues).[29] Touché sensors and conductive threads can be inserted or woven into everyday objects or even liquids. A Touché-enhanced doorknob can tell whether it being grasped or simply tapped, allowing doors to be programmed to lock or unlock using a gesture password. The liquid version of the sensor is reportedly so sensitive that it can detect whether the water surface is being poked with a single fingertip or whether an entire hand is being immersed. Touché sensors can even differentiate between individual users simply based on the unique impedance (electrical resistance) signature of each human body (like the electrical equivalent of a fingerprint). Touché-enhanced objects can also detect nearby movement, thus making them exquisitely sensitive to body gestures. Similar devices are now available for consumers: Google's Nest Hub Max detects gestures, so you can pause music or turn off an alarm simply by raising your hand—no direct contact required.

To illustrate the potential of swept frequency capacitive sensing, the inventor of Touché, Ivan Poupyrev, decided to hook up his sensors to living plants. His subsequent project for Disney was Botanicus Interacticus, which applied Touché technology to a variety of plants through a simple, noninvasive wire placed anywhere in the plant's soil. Botanicus Interacticus enables humans to interact expressively with plants, by rendering plants exquisitely sensitive to human gestures (whether touching, grasping, or sliding fingers along the plant) and tracking proximity between the human "user" and the plant. Says Poupyrev (who left Disney for Google's Advanced Technology and Projects lab):

> We envision a future where interactive devices are not manufactured but are living, growing organisms. . . . With Botanicus Interacticus technology, any living plant can be turned into a musical instrument, a game controller, a light switch, an elevator button, or a motion detector. Interactivity and responsiveness can be added to homes, streets, and even entire cities, unobtrusively and sustainably simply by growing plants.[30]

Poupyrev's work on Touché was inspired by the father of ubiquitous computing, Mark Weiser. A philosophy undergrad dropout who began coding on the side to pay his bills, Weiser eventually became the director of Xerox's famous PARC laboratory and pioneered the concept of disappearing computers. He believed that computers would one day be absorbed into everyday objects and aimed to shift computing out of devices and into our surroundings, as integral yet invisible technologies. As Weiser put it in a seminal article in *Scientific American*, "the most profound technologies are those that disappear."[31]

When Weiser wrote those words in 1991, this seemed like an improbable vision. The Cray C90 supercomputer, released the same year, was the size of a small walk-in closet and offered less storage capacity than today's high-end smartphones. Thirty years later, we've succeeded in miniaturizing computing power. But this is only one step toward ubiquitous computing. Getting rid of interfaces (no buttons to press, mice to move, or screens to swipe) still seems like an elusive goal. Completely new types of interactive technologies are required.

Until recently, most discussions of ubiquitous computing assumed that the new materials required to weave computers seamlessly into the fabric of our everyday lives would be artificial and mechanical. Much effort has been put into inventing new types of "smart materials," into which digital components have been embedded. E-textiles, for example, are created by weaving or 3D-printing waterproof microelectronic devices (like sensors or miniaturized semiconductors) directly into the fabric. Such "soft hardware" fabrics can be used as advanced communication devices, as smart clothing that monitors your heart rate while you exercise, or as medical devices. Some researchers hope that by endowing materials themselves with information-processing capabilities, all matter could one day be programmable.[32] Then computers could truly become ubiquitous, and we would have achieved Weiser's dream of making them "disappear": smart matter in the world around us would enable continuous data gathering about our bodies, our possessions, and our environments.

Although Weiser's vision for ubiquitous computing was digital, as we have discussed in this chapter, it is through biodigital innovation that computing will truly disappear. The boundaries between carbon-based life and digital data are becoming increasingly porous. The concept of the Internet of Living Things captures the notion that, in the future, our computers and digital devices might be entirely biological. Our digital technologies will be made from living things rather than dead things, and embedded in the living beings dwelling alongside us.

This book began with a discussion of digital technologies, and referred to the growth of the "tech stack" to a global scale. I have argued that Digital Earth technologies create the possibility for new types of environmental governance: digitally enabled, data-intensive, real-time, predictive, automated, spatially and temporally dynamic, collaborative, and (in the future) increasingly virtual and biodigital. But digital technologies, as we know them, are transient; they will one day be fossils no less curious than the antiquated tools revealed by the archaeologist. The eco-tech stack that I described in chapter 2 will seem clunky, even outdated, in the not-so-distant future. Gaia's Web is becoming digital, but will outgrow what we understand to be digital infrastructure like a snake sheds its skin, or a crab sheds it shell. What will be left? Computation already existed within the natural world, long before we discovered it; in the future, it will be re-embedded. What this means for our relationship with the environment is unclear, but we can already discern the deep challenge that this poses to our understanding of the distinction between humans and nature, and the relationship of humanity to other species and to our planet.

GREEN PROMETHEUS

Centuries from now, environmental historians might look back at our century and observe a slow but steady trend: a decline of anthropocentrism, the belief that humans are the most important and intelligent beings on the planet. Anthropocentrism has a complex lineage, but we can trace the concept back to Descartes, whose twin claims are still influential today: we think, therefore we are; and animals, lacking reason, are mere automata. Descartes celebrated human rationality and believed that humans were separate from—and superior to—nature. Cartesian views set the stage for contemporary anthropocentrism that has long justified the domination of nonhumans by humans.

Yet anthropocentrism is now under siege on multiple fronts: technological, environmental, and scientific. On the technological frontier, anthropocentrism is increasingly obsolete. The argument, in a nutshell, is this: as artificial intelligence develops, the uniqueness and agency of humans dwindles. As large swathes of the economy are automated, humans will be displaced, surplus. As AI surpasses human capacities in many domains, our claims to superior intelligence will no longer hold true. Anthropocentrism will become an outdated concept.

Environmentalists object to anthropocentrism too, but for a different reason. For environmentalists, anthropocentrism is morally objectionable because it has justified the ransacking of Earth by humanity. Moreover, the failure of the global community to address urgent global climate and biodiversity crises demonstrates that anthropocentrism is also ineffective, as it does not enable long-term human survival.

As discussed in my previous book *The Sounds of Life*, anthropocentrism is also being challenged by scientific research that demonstrates that animals, insects, and plants exhibit complex behaviors such as communication, learning, and memory. These findings are not new insights (Indigenous knowledge provides a long-standing perspective on these issues), but the weight of accumulated scientific evidence has opened up new debate within the scientific community, as many species display characteristics (such as complex communication) that Western science previously portrayed as the sole preserve of *Homo sapiens*. If characteristics of sentience and sapience, and perhaps even consciousness, are less uniquely human than we thought, then anthropocentrism may be both empirically incorrect and intellectually impoverished.

If the death of anthropocentrism—obsolete, morally objectionable, ineffective, and empirically incorrect—is looming, what will replace it? Digital innovation may enable us to move beyond anthropocentrism, but toward what? Digital biocentrism—which uses digital technologies to reconnect to nature and renew our relationships with other living beings—may provide a path forward. In chapter 6, I argued that digital biocentrism could allow new forms of environmental governance that incorporate nonhuman preferences and constrain human action. In chapter 7, I argued that digital biocentrism could also create new forms of property rights and legal personhood for nonhumans. And in chapter 8, I argued that digital biocentrism may enable new forms of sensing and empathizing with the lived experiences of nonhumans. Biodigital technologies enable nonhumans to engage in new types of political relationships: to assume legal personhood, have their own property rights, participate in environmental governance, and exercise political voice.

Digital innovation may enable new ways in which nonhumans can influence human decisions, by incorporating input from nonhumans into decision making. These technologies could, I argue, help create a new basis for new types of relationships with nonhumans that recognize and respect their autonomy and agency and enable political systems to evolve to recognize nonhumans as political subjects. With proper safeguards digital technologies could provide the mechanisms through which abstract "rights of nature" can become a reality, and nonhuman beings can assume legal standing through possessing rights and through property ownership. Digital Earth technologies could enable nonhumans to exercise political voice, analogous to the right to vote: a digitally enabled multispecies democracy.

But are these true collaborations, or do they simply instrumentalize nonhumans for the benefit of humans? It up to us to decide whether biodigital technologies will facilitate progressive politics or a new era of domination. But even this framing of the question is tricky. In the nineteenth century, colonialism was promoted with a savior ethos. In hindsight, the industrialization of natural commodity production in the nineteenth century, combined with colonialism and capitalism, was one of the main drivers of environmental degradation and human oppression worldwide. Often, the two went hand in hand. Are the animals, plants, and insects enrolled in creating our biodigital futures no more than technologically enhanced beasts of burden?

REWEAVING GAIA'S WEB

This book has traced the evolution of Gaia's Web through the lens of digital innovation. As a metaphor, Gaia's Web describes the Earth as simultaneously natural and technological, ecological and digital, human and nonhuman:

a merging of the Web of Life and the World Wide Web. Our future will be complicated by the blurring of boundaries between humans, nonhumans, and machines. Artificial intelligence and biohybrid engineering have accelerated our ability to merge biological and artificial worlds, both physically and cognitively. New organisms that fuse "hard" (artificial) and "wet" (natural) life intermingle natural and engineered ecologies. Biomimetic and biohybrid robotics, biological cybernetics, organismal engineering, bionic engineering—the strategies for interspecies mixing have proliferated far faster than our ability to regulate them. This has led to a new generation of biohybrid robots that blur the boundary between the biological and the synthetic. Through innovations such as these, cyborgs—part machine, part living organism—are being empowered to act as autonomous agents. Combining living organisms with digital technologies will create a set of new roles for nonhumans that we cannot fully anticipate. My argument is neither transhumanist—as I do not argue that intelligent machines should or will transcend humanity—nor ecomodernist—as I do not believe that technology will solve all of our environmental problems. Rather, I have argued for something more pragmatic: directing digital innovation to environmentally sustainable ends.

But is it that simple? By invoking technological solutions, critics argue, we may be mistaking the disease for the cure. Whether or not you think these technologies will solve our global environmental crises depends, in part, on your diagnosis of the causes. If climate change is the symptom, what is the underlying disease? Is it capitalism? Industrialization? Colonialism? All of the above?

Some argue that we have no time for debate: our house is literally burning down around us, and we should use any tools at our disposal to solve the climate emergency. Others argue that industrial capitalism has worsened resource extraction and environmental degradation; it is unlikely that surveillance eco-capitalism would be any different. These critics point to the well-known risks of the digital economy: undermining state sovereignty and democratic debate, concentrating economic and data-based power, producing surplus labor due to automation, threatening individual privacy. And they add to this list a set of urgent environmental risks: exceeding planetary boundaries due to digitally driven emissions- and resource-intensive growth, eco-surveillance, the privatization of environmental data, e-waste, and the ethical challenges posed by human enhancement and human-machine-organism hybrids and chimeras. There is no guarantee that digital technology will foster an ethics of reciprocity, reparation, repair, and regeneration. Why would powerful humans choose communion rather than dominion, kinship rather than ownership?

To answer this question, perhaps we might return again to the myth of Pandora, remembering that after the dangers and disasters are unleashed upon humanity, one deity—Elpis, goddess of hope—remains shut in the jar. Is this in order to preserve hope for humanity, or to preserve humanity from hope? In the former interpretation, hope is a wellspring, a source of inspiration and energy; in the latter interpretation, hope is a trap, false and misleading—the most perilous of the dangers facing humanity. Scientists are wiring up the planet with digital technologies, in hopes of mobilizing the tools of the Digital Age to solve the most pressing challenges of the Anthropocene, yet summoning the meshing and merging of digital and biological innovation in order to better manipulate organisms and ecosystems.

Is it rational or delusional to invoke hope at a moment like this? I have chosen to err on the side of cautious optimism, and explore ideas that imagine more abundant, inclusive futures. Of course, hope might seem like a foolish choice when facing an unprecedented existential threat. Throughout the book, I have tried to balance what Sheila Jasanoff calls the "can do" optimism of scientists and engineers with the "should do" skepticism of philosophers and ethicists. From this uneasy yet fertile middle ground, I argue that digital technologies are profoundly changing not only our relationships with the natural world but also our understanding of what it means to be human. If deployed with safeguards, digital technologies will help humanity connect and communicate with nonhumans, repair and regenerate the environment, and create a better basis for sharing our planet with our fellow Earthlings, cultivating habitability and sustainability in the places we collectively call home. Is this false hope? You decide.

PARABLE OF THE LONG NOW

Once upon a time, there was a clock designed to last 10,000 years. Its inventors hoped that the clock would encourage humanity to think more carefully about the future, so they christened their invention the Clock of the Long Now. The clock would tick once every hundred years, and chime just once every millennium.

The Clock of the Long Now was the most expensive time-keeping machine ever built, so the inventors asked the richest man in the world to pay for it. He agreed, but stipulated that the clock should be protected from all possible threats: rain and snow, vandals and looters, mold and pests. The clock was built deep inside a mountain, in a cave with many doors, far from cities or roads.

On top of the mountain lived bristlecone pines that were 5,000 years old, some of the most elderly beings in the world. The men were happy that their clock would outlive the trees. Centuries hence, they said, people would gather to see the clock chime on the millennium. At the ceremony, they would remember the benefactor. The Clock of the Long Now would conquer Chronos, the god of time.

Yet the inventors honored the wrong god.

Graying Chronos, an elderly man, is bound to the zodiac wheel, grinding monuments and memories into dust. The inventors forgot about the Horae, goddesses of the hours and seasons, guardians of the gates of Olympus, who once crowned Pandora with flowers. To honor the Horae, the inventors would have inserted a genetic barcode into the pine trees and their fungal networks, and encoded stories of human elders in the DNA of the *Komagataeibacter rhaeticus* bacterium in the soil.

The code of life does not decay, and passes information down through the generations, into Deep Time. The men chose, instead, to worship a machine in its tomb. The Clock of the Long Now is as quaint as a sundial, as advanced as a medieval robot. It is antiquated before even complete.

When might today's inventions seem like relics of the past?

Remembering what was forgotten, remembering the future, let us become the ancestors our descendants need us to be.

Acknowledgments

The concluding chapters of this book were written in 2023 while I was on sabbatical at the Radcliffe Institute for Advanced Study at Harvard University as the Matina S. Horner Distinguished Visiting Fellow. I'm deeply grateful for the support of Dr. Claudia Rizzini and the staff at the Institute, and to the wonderful colleagues in my cohort. The genesis of the book emerged during a sabbatical in 2015 at Stanford's Center for Advanced Study in the Behavioral Sciences, supported by a Leonore Annenberg and Wallis Annenberg Fellowship in Communication. The same year, I held a Cox Visiting Professorship at Stanford's School of Earth, Energy and Environmental Sciences. I'm grateful to my colleagues and hosts, including Dr. Rosemary Knight and Dr. Margaret Levi.

Many colleagues have shaped the ideas in these pages with generous and thoughtful commentary: Rahul Bhatia (independent writer, India), Jonathan Blake (Berggruen Institute), Jennifer Finney Boylan (Barnard College), Dirk Brinkman (independent writer), Jim Collins (Arizona State University), Bruno Correia (École Polytechnique Fédérale de Lausanne), Tsitsi Dangarembga (independent writer, Zimbabwe), Omar Dewachi (Rutgers University), Tawanna Dillahunt (University of Michigan), Jonathan Fink (Portland State University), Ebony Flowers (independent writer), Isabelle Galleymore (University of Birmingham), Anne Gorsuch (University of British Columbia), Rebecca Hall (independent scholar), Leila Harris (UBC), Nina Hewitt (UBC), John Huizinga (independent writer), Rosemary Knight (Stanford University), James Leape (Stanford

University), Kevin Leyton-Brown (UBC), Alan Mackworth (UBC), Asifa Majid (Oxford University), Louise Mandell (Mandell Pinder LLP), Yahya Modarres-Sadeghi (University of Massachusetts Amherst), Raymond Ng (UBC), Chris Reimer (UBC), Max Ritts (Clark University), Doug Robb (UBC), Joe Roman (University of Vermont), Stuart Rush (UBC), and Hong Yang (Bryant University). Philippe Le Billon (UBC) read numerous drafts and provided invaluable support throughout the process.

Various iterations of this book were discussed at presentations at Cambridge University, the Canadian Science Policy Centre, the Coalition for Digital Environmental Sustainability conference, Harvard University, the IEEE International Conference on Smart Computing, l'Université de Lyon, l'Université de Montréal, l'Université d'Ottawa, l'Université de Strasbourg, Oxford University, Stanford University, the University of Toronto, Ohio State University, the United Nations World Data Forum, the University of California at Santa Cruz, the University of California at Los Angeles, the University of Oslo, the University of Waterloo, and Wageningen University, as well as a TED talk and a keynote talk to the International Environmental Communication Association. Interviews were conducted with Dr. Dyhia Belhabib (Ecotrust Canada), Dr. Tanya Berger-Wolf (Ohio State University), Dr. Fei Fang (Carnegie Mellon University), and Dr. Manuela Veloso (head of AI Research, JPMorgan Chase & Co). Material from previously published articles in *Earth System Governance* (CC-BY-NC-ND) and *Global Environmental Change* (CC BY license) was referenced in various chapters:

Karen Bakker and Max Ritts. "Smart Earth: A Meta-Review and Implications for Environmental Governance." *Global Environmental Change* 52 (2018): 201–211.

Karen Bakker. "Smart Oceans: Artificial Intelligence and Marine Protected Area Governance." *Earth System Governance* 13 (2022): 100141.

Max Ritts and Karen Bakker. "New Forms: Anthropocene Festivals and Experimental Environmental Governance." *Environment and Planning E: Nature and Space* 5, no. 1 (2022): 125–145.

Research assistance for this book (and for the Smart Earth project) was ably provided at UBC by Ramit Brata Biswas, Amanda Chambers, Alycia Felli, Oliver Gadoury, Sophie Galloway, Joshua Green, Caroline Hanna, Anna Kavenay, Charlotte Michaels, Gabrielle Plowens, Evan Powers, Clare

Price, Eva Streitz, Adèle Therias, Bentley Tse, and Sophia Wilson. Research assistance at Harvard University was provided by Michaela Bibby and Maegan Jong. Their support was enabled by funding from the Harvard Radcliffe Institute for Advanced Study and the Social Sciences and Humanities Research Council of Canada.

Last but not least, warm thanks are due to my editor Beth Clevenger, to Stephanie Sakson for copy editing, and to the team at the MIT Press for their creative commitment to supporting critical debates about technology and environmental sustainability.

Notes

PART I

1. Octavia E. Butler, *Parable of the Sower* (New York: Four Walls Eight Windows, 1993), 3.

CHAPTER 1

1. The stories about orcas in chapter 1 are based on several sources: Jason Michael Colby, *Orca: How We Came to Know and Love the Ocean's Greatest Predator* (Oxford: Oxford University Press, 2018); Simone Cominelli et al., "Noise Exposure from Commercial Shipping for the Southern Resident Killer Whale Population," *Marine Pollution Bulletin* 136 (2018): 177–200; John K. B. Ford, *Marine Mammals of British Columbia* (Victoria: Royal BC Museum, 2014); Daniel Francis and Gil Hewlett, "They Shoot Orcas, Don't They?," *The Tyee*, May 13, 2008, https://thetyee.ca/Life/2008/05/13/ShootingOrcas/; Mark Leiren-Young, *The Killer Whale Who Changed the World* (Vancouver, BC: Greystone Books, 2016); Thomas P. Quinn and James P. Losee, "Diverse and Changing Use of the Salish Sea by Pacific Salmon, Trout, and Char," *Canadian Journal of Fisheries and Aquatic Sciences* 79, no. 6 (2022): 1003–1021; Mei Sato, Andrew W. Trites, and Stéphane Gauthier, "Southern Resident Killer Whales Encounter Higher Prey Densities Than Northern Resident Killer Whales during Summer," *Canadian Journal of Fisheries and Aquatic Sciences* 78, no. 11 (2021): 1732–1743; Samuel K. Wasser et al., "Population Growth Is Limited by Nutritional Impacts on Pregnancy Success in Endangered Southern Resident Killer Whales (*Orcinus orca*)," *PLOS One* 12, no. 6 (2017); Ed Yong, "What a Grieving Orca Tells Us," *Atlantic*, August 14, 2018, https://www.theatlantic.com/science/archive/2018/08/orca-family-grief/567470/.
2. Rena Priest, "A Captive Orca and a Chance for Our Redemption," *High Country News*, April 1, 2020, https://www.hcn.org/issues/52.4/indigenous-affairs-wildlife-a-captive-orca-and-a-chance-for-our-redemption; Rena Priest, "The People beneath the Waves," Microsoft Unlocked, accessed May 5, 2023, https://unlocked.microsoft.com/people-beneath-the-waves/.
3. R. E. Burham et al., "The Combined Use of Visual and Acoustic Data Collection Techniques for Winter Killer Whale (*Orcinus orca*) Observations," *Global Ecology and Conservation* 8

(2016): 24–30; Marine Randon, Michael Dowd, and Ruth Joy, "A Real-Time Data Assimilative Forecasting System for Animal Tracking," *Ecology* 103, no. 8 (2022): e3718; Ruth Joy et al., "Integrating Visual and Acoustic Observations to Build an 'Intelligent' Killer Whale Movement Forecast System," *The Journal of the Acoustical Society of America* 150, no. 4 (2021): A284.

4. Ruth Joy et al., "Potential Benefits of Vessel Slowdowns on Endangered Southern Resident Killer Whales," *Frontiers in Marine Science* 6 (2019): 344.

5. For more information on the use of digital technology to study orca sounds, and on the Smart Ocean initiative, see Martin Heesemann et al., "Ocean Networks Canada: From Geohazards Research Laboratories to Smart Ocean Systems," *Oceanography* 27, no. 2 (2014): 151–153; Kate Moran et al., "Canada's Internet-Connected Ocean," *Frontiers in Marine Science* 8 (2022); Jennifer B. Tennessen et al., "Hidden Markov Models Reveal Temporal Patterns and Sex Differences in Killer Whale Behavior," *Nature Scientific Reports* 9, no. 1 (2019a): 1–12; Jennifer B. Tennessen et al., "Kinematic Signatures of Prey Capture from Archival Tags Reveal Sex Differences in Killer Whale Foraging Activity," *Journal of Experimental Biology* 222, no. 3 (2019): jeb191874; Brianna M. Wright et al., "Behavioral Context of Echolocation and Prey-Handling Sounds Produced by Killer Whales (*Orcinus orca*) during Pursuit and Capture of Pacific Salmon (*Oncorhynchus* spp.)," *Marine Mammal Science* 37, no. 4 (2021): 1428–1453.

6. Julie Cattiau, "AI's Killer (Whale) App," *The Keyword* (blog), *Google*, January 28, 2020, https://www.blog.google/technology/ai/protecting-orcas/.

7. For a more detailed discussion of Digital Earth hardware, see Karen Bakker and Max Ritts, "Smart Earth: A Meta-Review and Implications for Environmental Governance," *Global Environmental Change* 52 (2018): 201–211.

8. W. G. Meikle and Niels Holst, "Application of Continuous Monitoring of Honeybee Colonies," *Apidologie* 46 (2015): 10–22.

9. For more information on digital tracking of whales, see Susan E. Parks et al., "Acoustic Crypsis in Communication by North Atlantic Right Whale Mother–Calf Pairs on the Calving Grounds," *Biology Letters* 15, no. 10 (2019): 20190485; Peter L. Tyack et al., "Extreme Diving of Beaked Whales," *Journal of Experimental Biology* 209, no. 21 (2006): 4238–4253.

10. Peter Bauer, Bjorn Stevens, and Wilco Hazeleger, "A Digital Twin of Earth for the Green Transition," *Nature Climate Change* 11, no. 2 (2021): 80–83; Paul Voosen, "Europe Builds 'Digital Twin' of Earth to Hone Climate Forecasts," *Science* 370, no. 6512 (2020): 16–17.

11. Matthew Ponsford, "Martin Wikelski Interview: Tracking Animals Reveals Their Sixth Sense," *New Scientist*, March 31, 2022, https://www.newscientist.com/article/mg25433800-900-martin-wikelski-interview-tracking-animals-reveals-their-sixth-sense/.

12. Jonathan S. Wright et al., "Rainforest-Initiated Wet Season Onset over the Southern Amazon," *Proceedings of the National Academy of Sciences* 114, no. 32 (2017): 8481–8486.

13. Will Steffen et al., "Planetary Boundaries: Guiding Human Development on a Changing Planet," *Science* 347, no. 6223 (2015): 1–11.

14. Lovelock and Margulis proposed that the atmosphere, created by living things, in turn creates hospitable living conditions for the organisms that produce the air that animals, including humans, breathe. Scientific debate raged over the question of whether Gaia was a mechanism or mere metaphor, and whether a planet—a nonliving organism—could be said to self-regulate. My view is that Gaia is not a mechanism, but is a useful metaphor to express the idea that life creates the conditions for life; to disrupt the atmosphere is to risk disrupting life on Earth.

15. Other concepts that may interest the reader are Donna Haraway's *Staying with the Trouble: Making Kin in the Chthulucene* (Durham, NC: Duke University Press, 2016); Bruno Latour's "Gaia 2.0: Could Humans Add Some Level of Self-Awareness to Earth's Self-Regulation?," *Science* 361, no. 6407 (2018): 1066–1068; and James Lovelock's *Novacene: The Coming Age of Hyperintelligence* (Cambridge, MA: MIT Press, 2019).

16. Suzanne Simard, *Finding the Mother Tree: Uncovering the Wisdom and Intelligence of the Forest* (London: Penguin Books, 2022).

17. Vaclav Smil, *Invention and Innovation: A Brief History of Hype and Failure* (Cambridge, MA: MIT Press, 2023).

18. Sheila Jasanoff, "The Dangerous Appeal of Tech," *MIT Technology Review* 124, no. 4 (July–August 2021): 16–17.

19. Joseph W. Day, "The Poet's Elpis and the Opening of Isthmian 8," *Transactions of the American Philological Association (1974–2014)* 121 (1991): 47–61; George Kazantzidis and Dimos Spatharas, *Hope in Ancient Literature, History and Art* (Berlin: Walter de Gruyter, 2018).

CHAPTER 2

1. Dyhia Belhabib, interview with author, Vancouver, October 2020. All subsequent quotes from Belhabib in this chapter are from this interview.

2. Dyhia Belhabib and Philippe Le Billon, "Fish Crimes in the Global Oceans," *Science Advances* 8, no. 12 (2022).

3. Food and Agriculture Organization of the United Nations, "Fishing Fleet," *FAO*, 2022, https://www.fao.org/3/cc0461en/online/sofia/2022/fishing-fleet.html.

4. Álvaro Enríquez-de-Salamanca, "Influence of Climate Change, Overfishing and COVID19 on Irregular Migration in West Africa," *Climate and Development* 15, no. 3 (2022): 1–14; Jessica H. Jönsson, "Overfishing, Social Problems, and Ecosocial Sustainability in Senegalese Fishing Communities," *Journal of Community Practice* 27, nos. 3–4 (2019): 213–230; U. Thara Srinivasan et al., "Food Security Implications of Global Marine Catch Losses Due to Overfishing," *Journal of Bioeconomics* 12 (2010): 183–200; Cornelia E. Nauen and Simona T. Boschetti, "Fisheries Crimes, Poverty and Food Insecurity," in *Routledge Handbook of Maritime Security* (Abingdon: Routledge, 2022), 239–249; Annisa Chand, "Transversal Criminality at Sea," *Nature Food* 3, no. 4 (2022): 242.

5. IPBES, *The Global Assessment Report on Biodiversity and Ecosystem Services: Summary for Policymakers* (Bonn: Intergovernmental Science-Policy Platform on Biodiversity and Ecosystem Services, 2019).
6. AIS was originally developed for collision avoidance and is mandated for seagoing vessels by the International Maritime Organization.
7. Elvira Poloczanska, "Keeping Watch on the Ocean," *Science* 359, no. 6378 (2018): 864–865; David A. Kroodsma et al., "Tracking the Global Footprint of Fisheries," *Science* 359, no. 6378 (2018): 904–908.
8. Cade Metz, "'Businesses Will Not Be Able to Hide': Spy Satellites May Give Edge from Above," *New York Times*, January 24, 2019, https://www.nytimes.com/2019/01/24/technology/satellites-artificial-intelligence.html.
9. Beste İşleyen, "Technology and Territorial Change in Conflict Settings: Migration Control in the Aegean Sea," *International Studies Quarterly* 65, no. 4 (2021): 1087–1096; Georgios Glouftsios and Panagiotis Loukinas, "Perceiving and Controlling Maritime Flows: Technology, Kinopolitics, and the Governmentalization of Vision," *International Political Sociology* 16, no. 3 (2022).
10. Gianpaolo Coro, Anton Ellenbroek, and Pasquale Pagano, "An Open Science Approach to Infer Fishing Activity Pressure on Stocks and Biodiversity from Vessel Tracking Data," *Ecological Informatics* 64 (2021): 101384; Lauren Drakopulos et al., "Making Global Oceans Governance In/visible with Smart Earth: The Case of Global Fishing Watch," *Environment and Planning: Nature and Space* 6, no. 2 (2022); A. John Woodill et al., "Ocean Seascapes Predict Distant-Water Fishing Vessel Incursions into Exclusive Economic Zones," *Fish and Fisheries* 22, no. 5 (2021): 899–910.
11. Wim C. Mullie, "Apparent Reduction of Illegal Trawler Fishing Effort in Ghana's Inshore Exclusive Zone 2012–2018 as Revealed by Publicly Available AIS Data," *Marine Policy* 108 (2019): 103623.
12. Hilde M. Toonen and Simon R. Bush, "The Digital Frontiers of Fisheries Governance," *Journal of Environmental Policy & Planning* 22, no. 1 (2020): 125–137.
13. Nadia Drake, "They Saw Earth from Space. Here's How It Changed Them," *National Geographic Magazine* 233, no. 3 (2018): 68–77.
14. Eldon C. Hall, *Journey to the Moon: The History of the Apollo Guidance Computer* (Reston, VA: American Institute of Aeronautics and Astronautics, 1996).
15. W. Jason Morgan, "Rises, Trenches, Great Faults, and Crustal Blocks," *Journal of Geophysical Research* 73, no. 6 (1967): 1959–1982.
16. Jennifer Gabrys, *Program Earth: Environmental Sensing Technology and the Making of a Computational Planet* (Minneapolis: University of Minnesota Press, 2016).
17. Dorion Sagan, *Lynn Margulis: The Life and Legacy of a Scientific Rebel* (White River Junction, VT: Chelsea Green, 2012); John Gribbin, *He Knew He Was Right: The Irrepressible Life of James Lovelock* (London: Penguin UK, 2009).

18. Michael A. Garrett, "Expanding World Views: Can SETI Expand Its Own Horizons and That of Big History Too?," in *Expanding Worldviews: Astrobiology, Big History and Cosmic Perspectives* (Cham: Springer International, 2021), 53–64.

19. Timothy M. Lenton and James E. Lovelock, "Daisyworld Revisited: Quantifying Biological Effects on Planetary Self-Regulation," *Tellus B: Chemical and Physical Meteorology* 53, no. 3 (2001): 288–305; Andrew J. Wood et al., "Daisyworld: A Review," *Reviews of Geophysics* 46, no. 1 (2008).

20. Benjamin H. Bratton, *The Stack: On Software and Sovereignty* (Cambridge, MA: MIT Press, 2016).

21. Paul Edwards, *A Vast Machine: Computer Models, Climate Data, and the Politics of Global Warming* (Cambridge, MA: MIT Press, 2010).

22. Dennis L. Cade, "US Spy Satellites Used to Drop Photos in 'Film Buckets' from Space for Airplanes to Catch in Mid-Air," *PetaPixel*, August 31, 2014, https://petapixel.com/2014/08/31/us-spy-satellites-used-drop-photos-film-buckets-space-airplanes-catch-mid-air/.

23. Sarah Scoles, "Now Entering Orbit: Tiny Lego-Like Modular Satellites," *Wired*, December 29, 2019, https://www.wired.com/story/now-entering-orbit-tiny-lego-like-modular-satellites/.

24. United Nations Office for Outer Space Affairs, "Annual Report 2018" (United Nations Office for Outer Space Affairs, 2018).

25. Michael F. Goodchild, "Citizens as Sensors: The World of Volunteered Geography," *GeoJournal* 69, no. 4 (2007): 211–221; Sarah Elwood and Agnieszka Leszczynski, "New Spatial Media, New Knowledge Politics," *Transactions of the Institute of British Geographers* 38, no. 4 (2013): 544–559; P. Y. (Yola) Georgiadou et al., "Citizen Sensors or Extreme Publics? Transparency and Accountability Interventions on the Mobile Geoweb," *International Journal of Digital Earth* 7, no. 7 (2014): 516–533.

26. Paul Jepson and Richard J. Ladle, "Nature Apps: Waiting for the Revolution," *Ambio* 44, no. 8 (2015): 827–832.

27. Walter Jetz et al., "Biological Earth Observation with Animal Sensors," *Trends in Ecology & Evolution* 37, no. 8 (2022): 719–724; Roland Kays et al., "Terrestrial Animal Tracking as an Eye on Life and Planet," *Science* 348, no. 6240 (2015): 1–10.

28. Annalisa Di Bernardino, Valeria Jennings, and Giacomo Dell'Omo, "Bird-Borne Samplers for Monitoring CO_2 and Atmospheric Physical Parameters," *Remote Sensing* 14, no. 19 (2022): 4876; Walter Jetz et al., "Biological Earth Observation with Animal Sensors," *Trends in Ecology & Evolution* 37, no. 8 (2022): 719–724; Roland Kays et al., "Terrestrial Animal Tracking as an Eye on Life and Planet," *Science* 348, no. 6240 (2015): 1–10; John C. Ryan et al., "Oceanic Giants Dance to Atmospheric Rhythms: Ephemeral Wind-Driven Resource Tracking by Blue Whales," *Ecology Letters* 25, no. 11 (2022): 2435–2447; Henri Weimerskirch et al., "Ocean Sentinel Albatrosses Locate Illegal Vessels and Provide the First Estimate of the Extent of Nondeclared Fishing," *Proceedings of the National Academy of Sciences* 117, no. 6 (2020): 3006–3014.

29. Roland Kays et al., "Terrestrial Animal Tracking as an Eye on Life and Planet," *Science* 348, no. 6240 (2015): 1–10.

30. Michel Foucault, *Security, Territory, Population: Lectures at the Collège de France, 1977–78* (New York: Picador, 2007); Shoshana Zuboff, *The Age of Surveillance Capitalism: The Fight for a Human Future at the New Frontier of Power* (New York: Public Affairs, 2019).

31. Lucas Joppa, "A Planetary Computer to Avert Environmental Disaster," *Scientific American*, September 19, 2019, https://www.scientificamerican.com/article/a-planetary-computer-to-avert-environmental-disaster/.

32. Brad Smith, "A Healthy Society Requires a Healthy Planet," *Official Microsoft Blog* (blog), *Microsoft*, April 15, 2020, https://blogs.microsoft.com/blog/2020/04/15/a-healthy-society-requires-a-healthy-planet/.

33. Amy L. Luers, "Planetary Intelligence for Sustainability in the Digital Age: Five Priorities," *One Earth* 4, no. 6 (2021): 772–775.

34. Zuboff, *The Age of Surveillance Capitalism*.

CHAPTER 3

1. Tanya Berger-Wolf, phone interview with author, May 2020. All subsequent quotes from Berger-Wolf in this chapter are from this interview.

2. Zaven Arzoumanian, Jason Holmberg, and Brad Norman, "An Astronomical Pattern-Matching Algorithm for Computer-Aided Identification of Whale Sharks *Rhincodon typus*," *Journal of Applied Ecology* 42, no. 6 (2005): 999–1011.

3. Bradley M. Norman et al., "Undersea Constellations," *BioScience* 67, no. 12 (2017): 1029–1043.

4. Annie Sneed, "Astronomy Tool Helps ID Sharks," *Scientific American*, July 19, 2018, https://www.scientificamerican.com/podcast/episode/astronomy-tool-helps-id-sharks/.

5. Jackie Snow, "The World's Animals Are Getting Their Very Own Facebook," *Fast Company*, June 22, 2018, https://www.fastcompany.com/40585495/the-worlds-animals-are-getting-their-very-own-facebook.

6. Bob Langert, "10 Minutes with Josh Henretig, Microsoft," *GreenBiz*, December 17, 2018, https://www.greenbiz.com/article/10-minutes-josh-henretig-microsoft.

7. Oliver Milman, "The Killing of Large Species Is Pushing Them towards Extinction, Study Finds," *Guardian*, February 6, 2019, https://www.theguardian.com/world/2019/feb/06/the-killing-of-large-species-is-pushing-them-towards-extinction-study-finds.

8. Brett R. Scheffers et al., "Global Wildlife Trade across the Tree of Life," *Science* 366, no. 6461 (2019): 71–76.

9. Wendy Annecke and Mmoto Masubelele, "A Review of the Impact of Militarisation: The Case of Rhino Poaching in Kruger National Park, South Africa," *Conservation and Society* 14,

no. 3 (2016): 195–204; Rosaleen Duffy, "War, by Conservation," *Geoforum* 69 (2016): 238–248; Rosaleen Duffy et al., "Why We Must Question the Militarisation of Conservation," *Biological Conservation* 232 (2019): 66–73.

10. Fei Fang, phone interview with author, December 2020.

11. National Science Foundation, "Outwitting Poachers with Artificial Intelligence," *Phys.org*, April 21, 2016, https://phys.org/news/2016-04-outwitting-poachers-artificial-intelligence.html.

12. Fei Fang et al., "PAWS: A Deployed Game-Theoretic Application to Combat Poaching," *AI Magazine* 38, no. 1 (2017): 23–36.

13. Fei Fang et al., *Artificial Intelligence and Conservation* (Cambridge: Cambridge University Press, 2019).

14. Jacob Kamminga et al., "Poaching Detection Technologies: A Survey," *Sensors* 18, no. 5 (2018): 1474; Jesús Jiménez López and Margarita Mulero-Pázmány, "Drones for Conservation in Protected Areas: Present and Future," *Drones* 3, no. 1 (2019): 10.

15. Clive Cookson, "Science v Poachers: How Tech Is Transforming Wildlife Conservation," *Financial Times*, November 27, 2019, https://www.ft.com/content/47edbf58-0c6f-11ea-bb52-34c8d9dc6d84.

16. Peter H. Wrege et al., "Acoustic Monitoring for Conservation in Tropical Forests: Examples from Forest Elephants," *Methods in Ecology and Evolution* 8, no. 10 (2017): 1292–1301.

17. Aakash Lamba et al., "Deep Learning for Environmental Conservation," *Current Biology* 29, no. 19 (2019): R977–R982; Oliver R. Wearn, Robin Freeman, and David M. P. Jacoby, "Responsible AI for Conservation," *Nature Machine Intelligence* 1, no. 2 (2019): 72–73.

18. Chris Sandbrook, "The Social Implications of Using Drones for Biodiversity Conservation," *Ambio* 44, suppl. 4 (2015): 636–647.

19. Steven J. Cooke et al., "Troubling Issues at the Frontier of Animal Tracking for Conservation and Management," *Conservation Biology* 31, no. 5 (2017): 1205–1207.

20. Jessica J. Meeuwig, Robert G. Harcourt, and Frederick G. Whoriskey, "When Science Places Threatened Species at Risk," *Conservation Letters* 8, no. 3 (2015): 151–152.

21. Jean Sebastien Finger and Aurelien Francillon, "Unprotected Geo-Localisation Data through ARGOS Satellite Signals: The Risk of Cyberpoaching," *Proceedings of the 13th ACM Conference on Security and Privacy in Wireless and Mobile Networks* (New York: Association for Computing Machinery, 2020): 356–357.

22. Evan A. Eskew et al., "United States Wildlife and Wildlife Product Imports from 2000–2014," *Scientific Data* 7, no. 1 (2020): 22; Sharon Guynup, Chris R. Shepherd, and Loretta Shepherd, "The True Costs of Wildlife Trafficking," *Georgetown Journal of International Affairs* 21 (2020): 28–37.

23. Jennah Green et al., "Risky Business: Live Non-CITES Wildlife UK Imports and the Potential for Infectious Diseases," *Animals* 10, no. 9 (2020): 1632.

24. Joseph R. Harrison, David L. Roberts, and Julio Hernandez-Castro, "Assessing the Extent and Nature of Wildlife Trade on the Dark Web," *Conservation Biology* 30, no. 4 (2016): 900–904; David L. Roberts and Julio Hernandez-Castro, "Bycatch and Illegal Wildlife Trade on the Dark Web," *Oryx* 51, no. 3 (2017): 393–394.

25. Yunrui Ji et al., "Assessment of Current Trade of Exotic Pets on the Internet in China," *Biodiversity Science* 28, no. 5 (2020): 644.

26. Caroline Sayuri Fukushima, Stefano Mammola, and Pedro Cardoso, "Global Wildlife Trade Permeates the Tree of Life," *Biological Conservation* 247 (2020): 108503; Brett R. Scheffers et al., "Global Wildlife Trade across the Tree of Life," *Science* 366, no. 6461 (2019): 71–76.

27. Elizabeth R. Beardsley, "Poachers with PCs: The United States' Potential Obligations and Ability to Enforce Endangered Wildlife Trading Prohibitions against Federal Traders Who Advertise on eBay," *UCLA Journal of Environmental Law & Policy* 25 (2006): 1.

28. Sara Alfino and David L. Roberts, "Code Word Usage in the Online Ivory Trade across Four European Union Member States," *Oryx* 54, no. 4 (2020): 494–498; Lydia M. Yeo, Rachel S. McCrea, and David L. Roberts, "A Novel Application of Mark-Recapture to Examine Behaviour Associated with the Online Trade in Elephant Ivory," *PeerJ* 5 (2017): e3048.

29. Enrico Di Minin et al., "Machine Learning for Tracking Illegal Wildlife Trade on Social Media," *Nature Ecology & Evolution* 2, no. 3 (2018): 406–407; Enrico Di Minin et al., "A Framework for Investigating Illegal Wildlife Trade on Social Media with Machine Learning," *Conservation Biology* 33, no. 1 (2019): 210.

30. Lauren Harrington, David Macdonald, and Neil D'Cruze, "Popularity of Pet Otters on YouTube: Evidence of an Emerging Trade Threat," *Nature Conservation* 36 (2019).

31. Tuuli Toivonen et al., "Social Media Data for Conservation Science: A Methodological Overview," *Biological Conservation* 233 (2019): 298–315.

32. Nathan Sing, "Google-Backed Project Is Collecting Millions of Wildlife Camera-Trap Images," *CNN*, January 13, 2020, https://www.cnn.com/2020/01/13/world/wildlife-insights-camera-trap-scn-intl-c2e/index.html.

33. Sandra Díaz et al., "Pervasive Human-Driven Decline of Life on Earth Points to the Need for Transformative Change," *Science* 366, no. 6471 (2019): eaax3100; Sandra Díaz et al., "Set Ambitious Goals for Biodiversity and Sustainability," *Science* 370, no. 6515 (2020): 411–413.

34. Carla Gomes et al., "Computational Sustainability: Computing for a Better World and a Sustainable Future," *Communications of the ACM* 62, no. 9 (2019): 56–65.

35. Devis Tuia et al., "Perspectives in Machine Learning for Wildlife Conservation," *Nature Communications* 13, no. 1 (2022): 792.

36. Luis F. Gonzalez et al., "Unmanned Aerial Vehicles (UAVs) and Artificial Intelligence Revolutionizing Wildlife Monitoring and Conservation," *Sensors* 16, no. 1 (2016): 97.

37. Larissa S. M. Sugai, "Pandemics and the Need for Automated Systems for Biodiversity Monitoring," *Journal of Wildlife Management* 84, no. 8 (2020): 1424–1426.

38. Ivan Jarić et al., "iEcology: Harnessing Large Online Resources to Generate Ecological Insights," *Trends in Ecology & Evolution* 35, no. 7 (2020a): 630–639.

39. Paul Jepson and Richard J. Ladle, "Nature Apps: Waiting for the Revolution," *Ambio* 44, no. 8 (2015): 827–832.

40. Paige West, James Igoe, and Dan Brockington, "Parks and Peoples: The Social Impact of Protected Areas," *Annual Reviews of Anthropology* 35 (2006): 251–277.

41. Peter Dauvergne, "The Globalization of Artificial Intelligence: Consequences for the Politics of Environmentalism," *Globalizations* 18, no. 2 (2021): 285–299.

42. Peter Corkeron, "The Militarization of Conservation: A Different Perspective," *Endangered Species Research* 50 (2023): 75–79.

43. Duffy, "War, by Conservation"; Rosaleen Duffy, *Security and Conservation: The Politics of the Illegal Wildlife Trade* (New Haven, CT: Yale University Press, 2022); Rosaleen V. Duffy and Dan Brockington, "Political Ecology of Security: Tackling the Illegal Wildlife Trade," *Journal of Political Ecology* 29, no. 1 (2022): 21–35.

44. Louise Amoore, *The Politics of Possibility: Risk and Security Beyond Probability* (Durham, NC: Duke University Press, 2013).

45. Duffy et al., "Why We Must Question the Militarisation of Conservation."

CHAPTER 4

1. Brad Smith, "Microsoft Will Be Carbon Negative by 2030," *Official Microsoft Blog*, January 16, 2020, https://blogs.microsoft.com/blog/2020/01/16/microsoft-will-be-carbon-negative-by-2030/.

2. Soyeon Bae et al., "Radar Vision in the Mapping of Forest Biodiversity from Space," *Nature Communications* 10, no. 1 (2019): 4757.

3. Larry Fink, "A Fundamental Reshaping of Finance," *BlackRock*, January 2020, https://www.blackrock.com/corporate/investor-relations/larry-fink-ceo-letter.

4. Yuzhong Zhang et al., "Quantifying Methane Emissions from the Largest Oil-Producing Basin in the United States from Space," *Science Advances* 6, no. 17 (2020): 1–9.

5. Robert B. Jackson, "The Integrity of Oil and Gas Wells," *Proceedings of the National Academy of Sciences* 111, no. 30 (2014): 10902–10903.

6. Arvind P. Ravikumar et al., "Single-Blind Inter-Comparison of Methane Detection Technologies: Results from the Stanford/EDF Mobile Monitoring Challenge," *Elementa: Science of the Anthropocene* 7 (2019).

7. Steven C. Wofsy and Steve Hamburg, "MethaneSAT: A New Observing Platform for High Resolution Measurements of Methane and Carbon Dioxide," *AGU Fall Meeting Abstracts* (2019): A53F-02.

8. Ramón A. Alvarez et al., "Assessment of Methane Emissions from the US Oil and Gas Supply Chain," *Science* 361, no. 6398 (2018): 186–188.

9. Jon Goldstein, "We Flew over 8,000 Oil and Gas Wells. Here's What We Found," *EDF Voices* (blog), *Environment Defense Fund*, April 21, 2016, https://www.edf.org/blog/2016/04/21/we-flew-over-8000-oil-and-gas-wells-heres-what-we-found.

10. Emmaline Atherton et al., "Mobile Measurement of Methane Emissions from Natural Gas Developments in Northeastern British Columbia, Canada," *Atmospheric Chemistry and Physics* 17, no. 20 (2017): 12405–12420.

11. Jonathan Mingle, "Methane Detectives: Can a Wave of New Technology Slash Natural Gas Leaks?," *Yale Environment 360*, October 31, 2019, https://e360.yale.edu/features/methane-detectives-can-a-wave-of-new-technology-slash-natural-gas-leaks.

12. Paul Tullis, "New Technology Claims to Pinpoint Even Small Methane Leaks from Space," *New York Times*, November 11, 2020, https://www.nytimes.com/2020/11/11/climate/methane-leaks-satellite-space.html.

13. Giovanni Bettini, Giovanna Gioli, and Romain Felli, "Clouded Skies: How Digital Technologies Could Reshape 'Loss and Damage' from Climate Change," *Wiley Interdisciplinary Reviews: Climate Change* 11, no. 4 (2020): e650.

14. Ovidiu Csillik et al., "Monitoring Tropical Forest Carbon Stocks and Emissions Using Planet Satellite Data," *Scientific Reports* 9, no. 1 (2019): 1–12.

15. Tara O'Shea, "Monitoring Wildfire Risk Using Space and AI," *Planet*, August 19, 2019, https://www.planet.com/pulse/monitoring-wildfire-risk-using-space-and-ai/.

16. Trevor Bell, Robert Briggs, Ralf Bachmayer, and Shuo Li, "Augmenting Inuit Knowledge for Safe Sea-Ice Travel: The SmartICE Information System," *2014 Oceans-St. John's* (2014): 1–9.

17. Heather Davis and Zoe Todd, "On the Importance of a Date, or Decolonizing the Anthropocene," *ACME: An International E-Journal for Critical Geographies* 16, no. 4 (2017): 1–20.

18. Alexander Koch et al., "Earth System Impacts of the European Arrival and Great Dying in the Americas after 1492," *Quaternary Science Reviews* 207 (2019): 13–36.

19. Will Steffen et al., "The Anthropocene: Conceptual and Historical Perspectives," *Philosophical Transactions of the Royal Society A: Mathematical, Physical and Engineering Sciences* 369, no. 1938 (2011): 842–867.

20. Candis Callison, *How Climate Change Comes to Matter: The Communal Life of Facts* (Durham, NC: Duke University Press, 2015).

21. Daniel R. Wildcat, "Introduction: Climate Change and Indigenous Peoples of the USA," in *Climate Change and Indigenous Peoples in the United States*, ed. Julie K. Maldonado, Benedict Colombi, and Rajul Pandya (Cham: Springer International, 2013), 1–7.

22. Kyle Powys Whyte, "Justice Forward: Tribes, Climate Adaptation and Responsibility," in *Climate Change and Indigenous Peoples in the United States*, ed. Julie K. Maldonado, Benedict Colombi, and Rajul Pandya (Cham: Springer, 2014), 9–22.

23. Shaun Ansell and Jennifer Koenig, "CyberTracker: An Integral Management Tool Used by Rangers in the Djelk Indigenous Protected Area, Central Arnhem Land, Australia," *Ecological Management and Restoration* 12, no. 1 (2011): 13–25.

24. Tahu Kukutai and John Taylor, eds., *Indigenous Data Sovereignty: Toward an Agenda* (Canberra: Australian National University Press, 2016).

25. Jon Swaine, "Two Google Searches 'Produce the Same CO_2 as Boiling a Kettle,'" *Telegraph*, January 11, 2009, https://www.telegraph.co.uk/technology/google/4217055/Two-Google-searches-produce-same-CO2-as-boiling-a-kettle.html.

26. Lynne H. Kaack et al., "Aligning Artificial Intelligence with Climate Change Mitigation," *Nature Climate Change* 12, no. 6 (2022): 518–527.

27. Brad Smith, "Microsoft Will Be Carbon Negative by 2030," *Official Microsoft Blog*, January 16, 2020, https://blogs.microsoft.com/blog/2020/01/16/microsoft-will-be-carbon-negative-by-2030/.

28. Nicola Jones, "How to Stop Data Centres from Gobbling Up the World's Electricity," *Nature* 561, no. 7722 (2018): 163–167; Laura U. Marks, Joseph Clark, Jason Livingston, Denise Oleksijczuk, and Lucas Hilderbrand, "Streaming Media's Environmental Impact," *Media + Environment* 2, no. 1 (2020): 1–17; Janine Morley, Kelly Widdicks, and Mike Hazas, "Digitalisation, Energy and Data Demand: The Impact of Internet Traffic on Overall and Peak Electricity Consumption," *Energy Research and Social Science* 38 (2018): 128–137.

29. Camilo Mora et al., "Bitcoin Emissions Alone Could Push Global Warming above 2°C," *Nature Climate Change* 8, no. 11 (2018): 931–933.

30. Elie Kapengut and Bruce Mizrach, "An Event Study of the Ethereum Transition to Proof-of-Stake," *arXiv preprint arXiv:2210.13655* (2022).

31. See, for example, Greenpeace, "Clicking Clean: Who Is Winning the Race to Build a Green Internet?," *Greenpeace*, January 2017, http://www.clickclean.org/international/en/.

32. Jeffrey Mantz, "Improvisational Economies: Coltan Production in the Eastern Congo," *Social Anthropology* 16, no. 1 (2008): 34–50; James H. Smith, "Tantalus in the Digital Age: Coltan Ore, Temporal Dispossession, and 'Movement' in the Eastern Democratic Republic of the Congo," *American Ethnologist* 38, no. 1 (2011): 17–35.

33. Thibault Cheisson and Eric J. Schelter, "Rare Earth Elements: Mendeleev's Bane, Modern Marvels," *Science* 363, no. 6426 (2019): 489–493.

34. Isaac Appiah et al., "The Impact of Information and Communication Technology (ICT) on Carbon Dioxide Emissions: Evidence from Heterogeneous ICT Countries," *Energy & Environment* (2022): 0958305X221118877; Ben Lahouel et al., "The Threshold Effects of ICT on CO_2 Emissions: Evidence from the MENA Countries," *Environmental Economics and Policy Studies* (2022): 1–21.

35. Johan Rockström et al., "A Roadmap for Rapid Decarbonization," *Science* 355, no. 6331 (2017): 1269–1271.

36. Tilman Santarius, Johanna Pohl, and Steffen Lange, "Digitalization and the Decoupling Debate: Can ICT Help to Reduce Environmental Impacts while the Economy Keeps Growing?," *Sustainability* 12, no. 18 (2020): 1–20.

37. Goodness C. Aye and Prosper Ebruvwiyo Edoja, "Effect of Economic Growth on CO_2 Emission in Developing Countries: Evidence from a Dynamic Panel Threshold Model," *Cogent Economics & Finance* 5, no. 1 (2017): 1379239.

38. Steffen Lange, Johanna Pohl, and Tilman Santarius, "Digitalization and Energy Consumption: Does ICT Reduce Energy Demand?," *Ecological Economics* 176 (2020): 1–14.

39. Annika Rieger, "Does ICT Result in Dematerialization? The Case of Europe, 2005–2017," *Environmental Sociology* 7, no. 1 (2020): 1–12.

40. Christopher Irrgang et al., "Towards Neural Earth System Modelling by Integrating Artificial Intelligence in Earth System Science," *Nature Machine Intelligence* 3, no. 8 (2021): 667–674; Markus Reichstein et al., "Deep Learning and Process Understanding for Data-Driven Earth System Science," *Nature* 566, no. 7743 (2019): 195–204.

41. Walter Leal Filho et al., "Deploying Artificial Intelligence for Climate Change Adaptation," *Technological Forecasting and Social Change* 180 (2022): 121662.

42. Anders Nordgren, "Artificial Intelligence and Climate Change: Ethical Issues," *Journal of Information, Communication and Ethics in Society* 21, no. 1 (2022).

43. Emma Strubell, Ananya Ganesh, and Andrew McCallum, "Energy and Policy Considerations for Deep Learning in NLP," *arXiv:1906.02243* (2019): 1–6.

44. Chris Stokel-Walker, "The Generative AI Race Has a Dirty Secret," *Wired*, October 2, 2023, https://www.wired.co.uk/article/the-generative-ai-search-race-has-a-dirty-secret.

45. George Caffentzis, *In Letters of Blood and Fire: Work, Machines, and the Crisis of Capitalism* (Oakland, CA: PM Press, 2012); Larry Lohmann, "Labour, Justice and the Mechanization of Interpretation," *Development* 62, nos. 1–4 (2019): 43–52.

46. Josh Cowls et al., "The AI Gambit: Leveraging Artificial Intelligence to Combat Climate Change—Opportunities, Challenges, and Recommendations," *AI & Society* (2021): 1–25.

47. David Rolnick et al., "Tackling Climate Change with Machine Learning," *arXiv:1906.05433* (2019): 1–111.

48. Lasse F. Wolff Anthony, Benjamin Kanding, and Raghavendra Selvan, "Carbontracker: Tracking and Predicting the Carbon Footprint of Training Deep Learning Models," *arXiv:2007.03051* (2020).

49. Mark Harris, "Artificial Intelligence and Decarbonization," *Anthropocene Magazine*, July 2017, http://www.anthropocenemagazine.org/AI/.

50. Nick Statt, "Google and DeepMind Are Using AI to Predict the Energy Output of Wind Farms," *Verge*, February 26, 2019, https://www.theverge.com/2019/2/26/18241632/google-deepmind-wind-farm-ai-machine-learning-green-energy-efficiency.

51. James Vincent, "Google Uses DeepMind AI to Cut Data Center Energy Bills," *Verge*, July 21, 2016, https://www.theverge.com/2016/7/21/12246258/google-deepmind-ai-data-center-cooling.

52. Amna Mughees et al., "Towards Energy Efficient 5G Networks Using Machine Learning: Taxonomy, Research Challenges, and Future Research Directions," *IEEE Access* 8 (2020): 187498–187522.

53. Mark Coeckelbergh, "AI for Climate: Freedom, Justice, and Other Ethical and Political Challenges," *AI and Ethics* 1 (2020): 67–72.

CHAPTER 5

1. Jan Zalasiewicz et al., "Scale and Diversity of the Physical Technosphere: A Geological Perspective," *Anthropocene Review* 4, no. 1 (2017): 9–22.

2. Karen Holmberg, "Landing on the Terrestrial Volcano," in *Critical Zones: The Science and Politics of Landing on Earth*, ed. B. Latour and P. Weibel (Cambridge, MA: MIT Press, 2020), 56–57.

3. Erle C. Ellis and Navin Ramankutty, "Putting People in the Map: Anthropogenic Biomes of the World," *Frontiers in Ecology and the Environment* 6, no. 8 (2008): 439–447.

4. Fridolin Krausmann et al., "Global Human Appropriation of Net Primary Production Doubled in the 20th Century," *Proceedings of the National Academy of Sciences* 110, no. 25 (2013): 10324–10329.

5. World Wildlife Fund, "Living Planet Report—2018: Aiming Higher," *WWF*, 2018, https://www.wwf.org.uk/sites/default/files/2018-10/wwfintl_livingplanet_full.pdf.

6. Jan Dönges, "All of Humanity Weighs Six Times as Much as All Wild Mammals," *Scientific American*, March 8, 2023, https://www.scientificamerican.com/article/all-of-humanity-weighs-six-times-as-much-as-all-wild-mammals/.

7. Will Steffen et al., "The Trajectory of the Anthropocene: The Great Acceleration," *Anthropocene Review* 2, no. 1 (2015): 81–98.

8. Global Commission on the Economy and the Climate, "Unlocking the Inclusive Growth Story of the 21st Century: Accelerating Climate Action in Urgent Times," August 2018, Global Commission on the Economy and the Climate, http://newclimateeconomy.report/2018/; Global Commission on the Economy and the Climate, "Press Release: Bold Climate Action Could Deliver US$26 Trillion to 2030, Finds Global Commission," September 5, 2018, Global Commission on the Economy and the Climate, https://newclimateeconomy.net/content/press-release-bold-climate-action-could-deliver-us26-trillion-2030-finds-global-commission.

9. Cameron Hepburn and Alex Bowen, "Prosperity with Growth: Economic Growth, Climate Change and Environmental Limits," in *Handbook on Energy and Climate Change* (Cheltenham: Edward Elgar, 2013), 617–638; Maria Kaika, Angelos Varvarousis, Federico

Demaria, and Hug March, "Urbanizing Degrowth: Five Steps towards a Radical Spatial Degrowth Agenda for Planning in the Face of Climate Emergency," *Urban Studies* (2023): 00420980231162234.

10. Nick Srnicek, *Platform Capitalism* (Cambridge: Polity Press, 2017); Soshana Zuboff, *The Age of Surveillance Capitalism* (New York: Public Affairs, 2019).

11. Bruno Basso and John Antle, "Precision Farming to Design Sustainable Agricultural Systems," *Nature Sustainability* 3, no. 4 (2020): 254–256.

12. Luis Ruiz-Garcia and Loredana Lunadei, "The Role of RFID in Agriculture: Applications, Limitations and Challenges," *Computers and Electronics in Agriculture* 79, no. 1 (2011): 42–50.

13. David C. Slaughter, Durham K. Giles, and Dennis J. Downey, "Autonomous Robotic Weed Control Systems: A Review," *Computers and Electronics in Agriculture* 61, no. 1 (2008): 63–78.

14. Andreas Kamilaris, Agusti Fonts, and Francesc X. Prenafeta-Boldú, "The Rise of Blockchain Technology in Agriculture and Food Supply Chains," *Trends in Food Science & Technology* 91 (2019): 640–652; Evagelos D. Lioutas and Chrysanthi Charatsari, "Big Data in Agriculture: Does the New Oil Lead to Sustainability?," *Geoforum* 109 (2020): 1–3; Sjaak Wolfert et al., "Big Data in Smart Farming: A Review," *Agricultural Systems* 153 (2017): 69–80.

15. John P. Fulton and Kaylee Port, "Precision Agriculture Data Management," in *Precision Agriculture Basics*, ed. D. Kent Shannon, David E. Clay, and Newell R. Kitchen (Madison, WI: American Society of Agronomy), 169–187.

16. Marco Springmann et al., "Options for Keeping the Food System within Environmental Limits," *Nature* 562, no. 7728 (2018): 519–525.

17. Robert E. Evenson and Douglas Gollin, "Assessing the Impact of the Green Revolution, 1960 to 2000," *Science* 300, no. 5620 (2003): 758–762; Prabhu L. Pingali, "Green Revolution: Impacts, Limits, and the Path Ahead," *Proceedings of the National Academy of Sciences* 109, no. 31 (2012): 12302–12308.

18. James R. Stevenson et al., "Green Revolution Research Saved an Estimated 18 to 27 Million Hectares from Being Brought into Agricultural Production," *Proceedings of the National Academy of Sciences* 110, no. 21 (2013): 8363–8368.

19. Laurens Klerkx, Emma Jakku, and Pierre Labarthe, "A Review of Social Science on Precision Farming, Smart Farming and Agriculture 4.0: New Contributions and a Future Research Agenda," *NJAS-Wageningen Journal of Life Sciences* 90 (2019): 1–17.

20. Sait M. Say et al., "Adoption of Precision Agriculture Technologies in Developed and Developing Countries," *Online Journal of Science and Technology* 8, no. 1 (2018): 7–15.

21. Sahra Svensson et al., "The Emerging 'Right to Repair' Legislation in the EU and the US," *Proceedings from Going Green–Care Innovation* (2018): 27–29.

22. Christopher M. Wathes et al., "Is Precision Livestock Farming an Engineer's Daydream or Nightmare, an Animal's Friend or Foe, and a Farmer's Panacea or Pitfall?," *Computers and Electronics in Agriculture* 64, no. 1 (2008): 2–10.

23. Heike Baumüller, "The Little We Know: An Exploratory Literature Review on the Utility of Mobile Phone–Enabled Services for Smallholder Farmers," *Journal of International Development* 30, no. 1 (2018): 134–154.

24. Sachin S. Kamble, Angappa Gunasekaran, and Rohit Sharma, "Modeling the Blockchain Enabled Traceability in Agriculture Supply Chain," *International Journal of Information Management* 52 (2020): 1–16.

25. Jennifer Clapp and Sarah-Louise Ruder, "Precision Technologies for Agriculture: Digital Farming, Gene-Edited Crops, and the Politics of Sustainability," *Global Environmental Politics* 20, no. 3 (2020): 49; Christopher Miles, "The Combine Will Tell the Truth: On Precision Agriculture and Algorithmic Rationality," *Big Data & Society* 6, no. 1 (2019): 1–12; Mark Shepherd et al., "Priorities for Science to Overcome Hurdles Thwarting the Full Promise of the 'Precision Farming' Revolution," *Journal of the Science of Food and Agriculture* 100, no. 14 (2020): 5083–5092.

26. Johan Colding and Stephan Barthel, "An Urban Ecology Critique of the 'Smart City' Model," *Journal of Cleaner Production* 164 (2017): 95–101.

27. Rob Kitchin, "The Ethics of Smart Cities and Urban Science," *Philosophical Transactions of the Royal Society A: Mathematical, Physical and Engineering Sciences* 374, no. 2083 (2016): 1–15.

28. Cities for Digital Rights, "Cities," *Cities for Digital Rights*, n.d., https://citiesfordigitalrights .org/cities.

29. Leyland Cecco, "'Surveillance Capitalism': Critic Urges Toronto to Abandon Smart City Projects," *The Guardian*, June 6, 2019, https://www.theguardian.com/cities/2019/jun/06 /toronto-smart-city-google-project-privacy-concerns.

30. Charlie Campbell, "'The Entire System Is Designed to Suppress Us': What the Chinese Surveillance State Means for the Rest of the World," *Time*, November 21, 2019, https://time .com/5735411/china-surveillance-privacy-issues/.

31. Joakim Krook and Leenard Baas, "Getting Serious about Mining the Technosphere: A Review of Recent Landfill Mining and Urban Mining Research," *Journal of Cleaner Production* 55 (2013): 1–9.

32. Sean Cubitt, *Finite Media: Environmental Implications of Digital Technologies* (Durham, NC: Duke University Press, 2016).

33. Peter Dauvergne, "Is Artificial Intelligence Greening Global Supply Chains? Exposing the Political Economy of Environmental Costs," *Review of International Political Economy* (2020): 1–23; Jonathan G. Koomey, H. Scott Matthews, and Eric Williams, "Smart Everything: Will Intelligent Systems Reduce Resource Use?," *Annual Review of Environment and Resources* 38 (2013): 311–343.

34. Anabel Marín and Daniel Goya, "Mining: The Dark Side of the Energy Transition," *Environmental Innovation and Societal Transitions* 41 (2021): 86–88.

35. Kevin Bales and Benjamin K. Sovacool, "From Forests to Factories: How Modern Slavery Deepens the Crisis of Climate Change," *Energy Research & Social Science* 77 (2021): 102096.

36. Benjamin K. Sovacool, "When Subterranean Slavery Supports Sustainability Transitions? Power, Patriarchy, and Child Labor in Artisanal Congolese Cobalt Mining," *Extractive Industries and Society* 8, no. 1 (2021): 271–293.

37. Bettina Engels, "African Anti-Mining Movements," in *The Wiley-Blackwell Encyclopedia of Social and Political Movements* (Hoboken, NJ: Wiley, 2013): 1–3; Benjamin K. Sovacool, "The Precarious Political Economy of Cobalt: Balancing Prosperity, Poverty, and Brutality in Artisanal and Industrial Mining in the Democratic Republic of the Congo," *Extractive Industries and Society* 6, no. 3 (2019): 915–939.

PART II

1. Frédérique Aït-Touati, "Arts of Inhabiting: Ancient and New Theaters of the World," in *Critical Zones: The Science and Politics of Landing on Earth*, ed. B. Latour and P. Weibel (Cambridge, MA: MIT Press, 2020), 432–438.

CHAPTER 6

1. The story of experimentation with acoustic deterrents is drawn from a number of historical sources, including William C. Cummings and Paul O. Thompson, "Gray Whales, *Eschrichtius robustus*, Avoid the Underwater Sounds of Killer Whales, *Orcinus orca*," *Fishery Bulletin* 69, no. 3 (1971): 525–530; and J. F. Fish and J. S. Vania, "Killer Whale, *Orcinus orca*, Sounds Repel White Whales, *Delphinapterus leucas*," *Fishery Bulletin* 69, no. 3 (1971): 531–535.

2. Nicola Jones, "Ocean Uproar: Saving Marine Life from a Barrage of Noise," *Nature* 568, no. 7752 (2019): 158–162.

3. R. Cotton Rockwood et al., "Modeling Whale Deaths from Vessel Strikes to Reduce the Risk of Fatality to Endangered Whales," *Frontiers in Marine Science* 8 (2021): 649890.

4. Megan Wood et al., "Near Real Time Passive Acoustic Monitoring in the Santa Barbara Channel," *Journal of the Acoustical Society of America* 148, no. 4 (2020): 2773.

5. Eliza Oldach et al., "Managed and Unmanaged Whale Mortality in the California Current Ecosystem," *Marine Policy* 140 (2022): 105039.

6. Kimberley T. A. Davies et al., "Mass Human-Caused Mortality Spurs Federal Action to Protect Endangered North Atlantic Right Whales in Canada," *Marine Policy 104* (2019): 157–162.

7. Delphine Durette-Morinet et al., "Passive Acoustic Monitoring Predicts Daily Variation in North Atlantic Right Whale Presence and Relative Abundance in Roseway Basin, Canada," *Marine Mammal Science* 35, no. 4 (2019): 1280–1303; Hansen Johnson, Daniel Morrison,

and Christopher Taggart, "WhaleMap: A Tool to Collate and Display Whale Survey Results in Near Real-Time," *Journal of Open Source Software* 6, no. 62 (2021): 3094.

8. Darren Incorvaia, "Recreational Fishing Industry Ranks the Safety of Right Whales below Profit," *Hakai Magazine*, May 9, 2023, https://hakaimagazine.com/features/recreational-fishing-industry-ranks-the-safety-of-right-whales-below-profit/.

9. Dipesh Chakrabarty, "The Planet Is a Political Orphan," *NOEMA*, February 3, 2022, https://www.noemamag.com/the-planet-is-a-political-orphan/.

10. Ralph Chami et al., "Nature's Solution to Climate Change: A Strategy to Protect Whales Can Limit Greenhouse Gases and Global Warming," *Finance & Development* 56, no. 4 (2019).

11. The discussion in this chapter draws on material previously published in Karen Bakker, "Smart Oceans: Artificial Intelligence and Marine Protected Area Governance," *Earth System Governance* 13 (2022): 100141.

12. Hans-Otto Portner et al., eds., "Summary for Policymakers," in *Special Report on the Ocean and Cryosphere in a Changing Climate* (Geneva: IPCC, 2019).

13. Lauren H. McWhinnie et al., "Vessel Traffic in the Canadian Arctic: Management Solutions for Minimizing Impacts on Whales in a Changing Northern Region," *Ocean & Coastal Management* 160 (2018): 1–17.

14. Alistair J. Hobday, Sara M. Maxwell, Julie Forgie, Jan McDonald, Marta Danby, Katy Seto, Helen Bailey, et al., "Dynamic Ocean Management: Integrating Scientific and Technological Capacity with Law, Policy, and Management," *Stanford Environmental Law Journal* 33, no. 125 (March 2013); Sara M. Maxwell, Elliott L. Hazen, Rebecca L. Lewison, Daniel C. Dunn, Helen Bailey, Steven J. Bograd, Dana K. Briscoe, et al., "Dynamic Ocean Management: Defining and Conceptualizing Real-Time Management of the Ocean," *Marine Policy* 58 (2015): 42–50; Sara M. Maxwell, Kristina M. Gjerde, Melinda G. Conners, and Larry B. Crowder, "Mobile Protected Areas for Biodiversity on the High Seas," *Science* 367, no. 6475 (2020): 252–254.

15. Jorge E. Corredor, "Platforms for Coastal Ocean Observing," in *Coastal Ocean Observing* (Cham: Springer, 2018), 67–84.

16. CSIRO, "Tagging Southern Bluefin Tuna in the Great Australian Bight," CSIRO: Australia's National Science Agency, 2020, https://www.csiro.au/en/research/animals/Fisheries/Tagging-fish; J. Paige Eveson, Alistair J. Hobday, Jason R. Hartog, Claire M. Spillman, and Kirsten M. Rough, "Seasonal Forecasting of Tuna Habitat in the Great Australian Bight," *Fisheries Research* 170 (2015): 39–49; Mark R. Payne, Alistair J. Hobday, Brian R. MacKenzie, Desiree Tommasi, Danielle P. Dempsey, Sascha M. Fässler, Alan C. Haynie, et al., "Lessons from the First Generation of Marine Ecological Forecast Products," *Frontiers in Marine Science* 4 (2017).

17. Australian Fisheries Management Authority, "Southern Bluefin Tuna," Australian Fisheries Management Authority—Australian Government, March 13, 2014, https://www.afma.gov.au/species/southern-bluefin-tuna; Stefano B. Longo, "Global Sushi: The Political Economy

of the Mediterranean Bluefin Tuna Fishery in the Modern Era," *Journal of World-Systems Research* (2011): 403–427.

18. Eveson et al., "Seasonal Forecasting of Tuna Habitat"; Jason R. Hartog and Alistair J. Hobday, "Case Study 8: Dynamic Spatial Management in an Australian Tuna Fishery," in *Biodiversity and Climate Change* (New Haven, CT: Yale University Press, 2019), 263–265; Alistair J. Hobday et al., "A Framework for Combining Seasonal Forecasts and Climate Projections to Aid Risk Management for Fisheries and Aquaculture," *Frontiers in Marine Science* 5 (2018): 137.

19. Alistair J. Hobday et al., "Dynamic Ocean Management: Integrating Scientific and Technological Capacity with Law, Policy, and Management," *Stanford Environmental Law Journal* 33 (2013): 125; William K. Oestreich, Melissa S. Chapman, and Larry B. Crowder, "A Comparative Analysis of Dynamic Management in Marine and Terrestrial Systems," *Frontiers in Ecology and the Environment* 18, no. 9 (2020): 496–504; Mark D. Reynolds et al., "Dynamic Conservation for Migratory Species," *Science Advances* 3, no. 8 (2017).

20. Maite Pons et al., "Trade-Offs between Bycatch and Target Catches in Static versus Dynamic Fishery Closures," *Proceedings of the National Academy of Sciences* 119, no. 4 (2022): e2114508119.

21. M. E. Gilmour et al., "Evaluation of MPA Designs That Protect Highly Mobile Megafauna Now and under Climate Change Scenarios," *Global Ecology and Conservation* 35 (2022): e02070.

22. Rahul Sharma, "Potential Impacts of Deep-Sea Mining on Ecosystems," in *Oxford Research Encyclopedia of Environmental Science* (Oxford: Oxford University Press, 2020).

23. Alan Boyle, "Law of the Sea Perspectives on Climate Change," in *The 1982 Law of the Sea Convention at 30: Successes, Challenges and New Agendas*, ed. David Freestone (Leiden: Martinus Nijhoff, 2013), 157–164; James Harrison, "Litigation under the United Nations Convention on the Law of the Sea: Opportunities to Support and Supplement the Climate Change Regime," in *Climate Change Litigation: Global Perspectives*, ed. Ivano Alogna, Christine Bakker, and Jean-Pierre Gauci (Leiden: Brill Nijhoff, 2021), 415–432; Seokwoo Lee and Lowell Bastista, "Part XII of the United Nations Convention on the Law of the Sea and the Duty to Mitigate against Climate Change: Making Out a Claim, Causation, and Related Issues," *Ecology Law Quartarly* 45, no. 1 (2018).

24. Morgan E. Visalli, Benjamin D. Best, Reniel B. Cabral, William W. L. Cheung, Nichola A. Clark, Cristina Garilao, Kristin Kaschner, et al., "Data-Driven Approach for Highlighting Priority Areas for Protection in Marine Areas beyond National Jurisdiction," *Marine Policy* 122 (2020): 103927.

25. Ina Tessnow-von Wysocki and Alice B. Vadrot, "The Voice of Science on Marine Biodiversity Negotiations: A Systematic Literature Review," *Frontiers in Marine Science* 7 (2020).

26. Elizabeth M. De Santo, Áslaug Ásgeirsdóttir, Ana Flavia Barros-Platiau, Frank Biermann, John Dryzek, Leandra Regina Gonçalves, Rakhyun E. Kim, et al., "Protecting Biodiversity in

Areas beyond National Jurisdiction: An Earth System Governance Perspective," *Earth System Governance* 2 (2019): 100029.

27. Daniel C. Dunn et al., "Empowering High Seas Governance with Satellite Vessel Tracking Data," *Fish and Fisheries* 19, no. 4 (2018): 729–739.

28. Elizabeth M. De Santo, "Implementation Challenges of Area-Based Management Tools (ABMTs) for Biodiversity beyond National Jurisdiction (BBNJ)," *Marine Policy* 97 (2018): 34–43.

29. Achille Mbembe, "How to Develop a Planetary Consciousness," *NOEMA*, January 11, 2022, https://www.noemamag.com/how-to-develop-a-planetary-consciousness/.

30. Bruno Latour, *We Have Never Been Modern* (Cambridge, MA: Harvard University Press, 1993).

31. Michel Serres, *The Natural Contract* (Ann Arbor: University of Michigan Press, 1995); Michel Serres, *The Parasite* (Minneapolis: University of Minnesota Press, 2013).

32. Bruno Latour, *Politics of Nature* (Cambridge, MA: Harvard University Press, 2004).

33. Related concepts include Earth System Law and a global environmental constitution for the Anthropocene. See Louis J. Kotzé and Rakhyun E. Kim., "Earth System Law: The Juridical Dimensions of Earth System Governance," *Earth System Governance* 1 (2019): 1–12.

34. Karen Bakker, *The Sounds of Life* (Princeton, NJ: Princeton University Press, 2022).

35. Thomas D. Seeley, *Honeybee Democracy* (Princeton, NJ: Princeton University Press, 2010).

CHAPTER 7

1. Sarah Vanuxem, "Freedom through Easements," in *Critical Zones: The Science and Politics of Landing on Earth*, ed. Bruno Latour and Peter Weibel (Cambridge, MA: MIT Press, 2020), 240–246.

2. Nelly Mäekivi, "Freedom in Captivity: Managing Zoo Animals According to the 'Five Freedoms,'" *Biosemiotics* 11, no. 1 (2018): 7–25; Steven P. McCulloch, "A Critique of FAWC's Five Freedoms as a Framework for the Analysis of Animal Welfare," *Journal of Agricultural and Environmental Ethics* 26, no. 5 (2012): 959–975; Clare McCausland, "The Five Freedoms of Animal Welfare Are Rights," *Journal of Agricultural and Environmental Ethics* 27, no. 4 (2014): 649–662.

3. Hope M. Babcock, "A Brook with Legal Rights: The Rights of Nature in Court," *Ecology Law Quarterly* 43 (2016): 1–52.

4. Catherine J. Iorns Magallanes, "From Rights to Responsibilities Using Legal Personhood and Guardianship for Rivers," in *ResponsAbility: Law and Governance for Living Well with the Earth*, ed. Betsan Martin, Linda Te Aho, and Maria Humphries-Kil (London: Routledge, 2018), 216–239.

5. Liz Charpleix, "The Whanganui River as Te Awa Tupua: Place-Based Law in a Legally Pluralistic Society," *Geographical Journal* 184, no. 1 (2018): 19–30; Christopher Rodgers, "A New Approach to Protecting Ecosystems: The Te Awa Tupua (Whanganui River Claims Settlement) Act 2017," *Environmental Law Review* 19, no. 4 (2017): 266–279.

6. John Burrows, *Drawing Out Law: A Spirit's Guide* (Toronto: University of Toronto Press, 2010); Val Napoleon, "Thinking about Indigenous Legal Orders," in *Dialogues on Human Rights and Legal Pluralism*, ed. René Provost and Colleen Sheppard (Dordrecht: Springer, 2013), 229–245.

7. Aimée Craft, *Breathing Life into the Stone Fort Treaty* (Vancouver: UBC Press, 2013).

8. Kelly Swing et al., "Outcomes of Ecuador's Rights of Nature for Nature's Sake," *Advances in Environmental and Engineering Research* 3, no. 3 (2022): 1–20.

9. Peter Howson, "Tackling Climate Change with Blockchain," *Nature Climate Change* 9, no. 9 (2019): 644–645; Peter Howson, "Distributed Degrowth Technology: Challenges for Blockchain beyond the Green Economy," *Ecological Economics* 184 (2021): 1–7; A. Stuitt, D. Brockington, and E. Corbera, "Smart, Commodified and Encoded: Blockchain Technology for Environmental Sustainability and Nature Conservation," *Conservation and Society* 20, no. 1 (2022): 12–23.

10. Guillaume Chapron, "The Environment Needs Cryptogovernance," *Nature* 545, no. 7655 (2017): 403–405.

11. Larry Lohmann, "Blockchain Machines, Earth Beings and the Labour of Trust," *The Corner House* (2020), https://www.researchgate.net/publication/333319658_Blockchain_Machines_Earth_Beings_and_the_Labour_of_Trust.

12. I have previously written about terra0 in the following article: Max Ritts and Karen Bakker, "New Forms: Anthropocene Festivals and Experimental Environmental Governance," *Environment and Planning E: Nature and Space* 5, no. 1 (2022): 125–145.

13. Jonathan Ledgard, "Interspecies Money," in *Breakthrough: The Promise of Frontier Technologies for Sustainable Development*, ed. Homi Kharas, John W. McArthur, and Izumi Ohno (Washington, DC: Brookings Institution Press, 2022), 77–102.

14. Matthew A. Zook and Joe Blankenship, "New Spaces of Disruption? The Failures of Bitcoin and the Rhetorical Power of Algorithmic Governance," *Geoforum* 96 (2018): 248–255; Jim Thatcher, David O'Sullivan, and Dillon Mahmoudi, "Data Colonialism through Accumulation by Dispossession: New Metaphors for Daily Data," *Environment and Planning D: Society and Space* 34, no. 6 (2016): 990–1006.

15. Nicholas Blomley, "Law, Property, and the Geography of Violence: The Frontier, the Survey, and the Grid," *Annals of the Association of American Geographers* 93, no. 1 (2003): 121–141.

16. Jennifer Gabrys, "Smart Forests and Data Practices: From the Internet of Trees to Planetary Governance," *Big Data and Society* 7, no. 1 (2020): 1–10.

17. Robert Herian, *Regulating Blockchain: Critical Perspectives in Law and Technology* (New York: Routledge, 2019).

18. Pierce Greenberg and Dylan Bugden, "Energy Consumption Boomtowns in the United States: Community Responses to a Cryptocurrency Boom," *Energy Research & Social Science* 50 (2019): 162–167; Christophe Schinckus, "The Good, the Bad and the Ugly: An Overview of the Sustainability of Blockchain Technology," *Energy Research & Social Science* 69 (2020): 101614.

19. Filipe Calvão and Matthew Archer, "Digital Extraction: Blockchain Traceability in Mineral Supply Chains," *Political Geography* 87 (2021): 102381; Peter Howson, "Distributed Degrowth Technology: Challenges for Blockchain beyond the Green Economy," *Ecological Economics* 184 (2021): 1–7.

20. Robert van den Hoven Van Genderen, "Do We Need New Legal Personhood in the Age of Robots and AI?," in *Robotics, AI and the Future of Law*, ed. Marcelo Corrales Compagnucci, Mark Fenwick, and Nikolaus Forgó (Singapore: Springer, 2018), 15–55.

21. Sue Donaldson and Will Kymlicka, *Zoopolis: A Political Theory of Animal Rights* (Oxford: Oxford University Press, 2011); Christopher Stone, "Should Trees Have Standing? Toward Legal Rights for Natural Objects," in *International Environmental Law*, ed. Paula M. Pevato (Oxfordshire: Routledge, 1972), 450–501.

22. Joshua C. Gellers, *Rights for Robots: Artificial Intelligence, Animal and Environmental Law* (London: Routledge, 2020).

23. Visa A. J. Kurki and Tomasz Pietrzykowski, eds., *Legal Personhood: Animals, Artificial Intelligence and the Unborn* (New York: Springer, 2017).

24. David A. Mindell, *Our Robots, Ourselves: Robotics and the Myths of Autonomy* (New York: Penguin Random House, 2015).

25. Gary Marcus and Ernest Davis, *Rebooting AI: Building Artificial Intelligence We Can Trust* (New York: Vintage Books, 2019).

26. Manuela Veloso, interview with author, November 2020.

27. Gary Marcus, "The Next Decade in AI: Four Steps towards Robust Artificial Intelligence," *arXiv:2002.06177* (2020).

28. Veloso, interview.

29. Donna Haraway, *Staying with the Trouble: Making Kin in the Chthulucene* (Durham, NC: Duke University Press, 2016); Donna Haraway, "Symbiogenesis, Sympoiesis, and Art Science Activisms for Staying with the Trouble," in *Arts of Living on a Damaged Planet: Ghosts and Monsters of the Anthropocene*, ed. Anna Lowenhaupt Tsing, Heather Anne Swanson, Elaine Gan, and Nils Bubandt (Minneapolis: University of Minnesota Press, 2017), M25–M50.

CHAPTER 8

1. Fernanda Herrera et al., "Building Long-Term Empathy: A Large-Scale Comparison of Traditional and Virtual Reality Perspective-Taking," *PLOS One* 13, no. 10 (2018): 1–37.

2. Sun Joo Ahn, Jeremy N. Bailenson, and Dooyeon Park, "Short- and Long-Term Effects of Embodied Experiences in Immersive Virtual Environments on Environmental Locus of Control and Behavior," *Computers in Human Behavior* 39 (2014): 235–245.

3. Rob Jordan, "Virtual Reality Could Serve as Powerful Environmental Education Tool, According to Stanford Researchers," *Stanford News*, November 30, 2018, https://news.stanford.edu/2018/11/30/virtual-reality-aids-environmental-education/.

4. Geraldine Fauville et al., "Participatory Research on Using Virtual Reality to Teach Ocean Acidification: A Study in the Marine Education Community," *Environmental Education Research* 27, no. 2 (2020): 1–25; David M. Markowitz et al., "Immersive Virtual Reality Field Trips Facilitate Learning about Climate Change," *Frontiers in Psychology* 9 (2018): 1–20.

5. Rob Jordan, "Virtual Reality Could Serve as Powerful Environmental Education Tool, Study Says," *Curriculum Review* 58, no. 7 (2019): 7.

6. Sun Joo Ahn, Amanda Minh Tran Le, and Jeremy Bailenson, "The Effect of Embodied Experiences on Self-Other Merging, Attitude, and Helping Behavior," *Media Psychology* 16, no. 1 (2013): 7–38; Nicola S. Schutte and Emma J. Stilinović, "Facilitating Empathy through Virtual Reality," *Motivation and Emotion* 41, no. 6 (2017): 708–712.

7. Austin van Loon et al., "Virtual Reality Perspective-Taking Increases Cognitive Empathy for Specific Others," *PLOS One* 13, no. 8 (2018): 1–19.

8. Nick Yee and Jeremy Bailenson, "The Proteus Effect: The Effect of Transformed Self-Representation on Behavior," *Human Communication Research* 33, no. 3 (2007): 271–290.

9. Marc Bekoff, *Minding Animals: Awareness, Emotions, and Heart* (Oxford University Press on Demand, 2002); Frans B. M. De Waal and Stephanie D. Preston, "Mammalian Empathy: Behavioural Manifestations and Neural Basis," *Nature Reviews Neuroscience* 18, no. 8 (2017): 498–509.

10. Nate Dolensek, Daniel A. Gehrlach, Alexandra S. Klein, and Nadine Gogolla, "Facial Expressions of Emotion States and Their Neuronal Correlates in Mice," *Science* 368, no. 6486 (2020): 89–94.

11. Joseph Henrich, Steven J. Heine, and Ara Norenzayan, "The Weirdest People in the World?," *Behavioral and Brain Sciences* 33, nos. 2–3 (2010): 61–83.

12. Jeremy Bailenson et al., "A Bird's Eye View: Biological Categorization and Reasoning within and across Cultures," *Cognition* 84, no. 1 (2002): 1–53.

13. Katherine M. Nelson, Eva Anggraini, and Achim Schlüter, "Virtual Reality as a Tool for Environmental Conservation and Fundraising," *PLOS One* 15, no. 4 (2020): 1–21.

14. Zeya He, Laurie Wu, and Xiang Robert Li, "When Art Meets Tech: The Role of Augmented Reality in Enhancing Museum Experiences and Purchase Intentions," *Tourism Management* 68 (2018): 127–139.

15. Sun Joo Ahn et al., "Experiencing Nature: Embodying Animals in Immersive Virtual Environments Increases Inclusion of Nature in Self and Involvement with Nature," *Journal of Computer-Mediated Communication* 21, no. 6 (2016): 399–419.

16. Kevin Kelly, "AR Will Spark the Next Big Tech Platform—Call It Mirrorworld," *Wired*, February 12, 2019, https://www.wired.com/story/mirrorworld-ar-next-big-tech-platform/.

17. Frans de Waal, *Are We Smart Enough to Know How Smart Animals Are?* (New York: W. W. Norton, 2016).

18. Adi Robertson, "What's Left of Magic Leap?," *Verge*, June 16, 2020, https://www.theverge.com/2020/6/16/21274638/magic-leap-app-store-partnerships-update.

19. Bram Büscher, "Nature 2.0: Exploring and Theorizing the Links between New Media and Nature Conservation," *New Media and Society* 18, no. 5 (2016): 726–743; Bram Büscher, "Conservation and Development 2.0: Intensifications and Disjunctures in the Politics of Online 'Do-Good' Platforms," *Geoforum* 79 (2017): 163–173; Bram Büscher, Stasja Koot, and Ingrid L. Nelson, "Introduction. Nature 2.0: New Media, Online Activism and the Cyberpolitics of Environmental Conservation," *Geoforum* 100, no. 79 (2017): 111–113; Bram Büscher and Jim Igoe, "'Prosuming' Conservation? Web 2.0, Nature and the Intensification of Value-Producing Labour in Late Capitalism," *Journal of Consumer Culture* 13, no. 3 (2013): 283–305.

20. Robert Fletcher, "Gaming Conservation: Nature 2.0 Confronts Nature-Deficit Disorder," *Geoforum* 79 (2017): 153–162.

21. Richard Louv, *Last Child in the Woods: Saving Our Children from Nature-Deficit Disorder* (Chapel Hill, NC: Algonquin Books, 2013).

22. Peter H. Kahn, Rachel L. Severson, and Jolina H. Ruckert, "Technological Nature—and the Problem When Good Enough Becomes Good," in *New Visions of Nature: Complexity and Authenticity*, ed. Martin A. M. Drenthen, F. W. Jozef Keulartz, and James Proctor (Dordrecht: Springer, 2009), 21–40.

23. Peter H. Kahn, *Technological Nature: Adaptation and the Future of Human Life* (Cambridge, MA: MIT Press, 2011).

24. David Abram, *The Spell of the Sensuous: Perception and Language in a More-Than-Human World* (New York: Vintage, 2012).

25. Suzanne Simard, *Finding the Mother Tree: Uncovering the Wisdom and Intelligence of the Forest* (London: Penguin UK, 2021).

26. Robin Kimmerer, *Braiding Sweetgrass: Indigenous Wisdom, Scientific Knowledge and the Teachings of Plants* (Minneapolis, MN: Milkweed Editions, 2013).

27. Gabrielle Walker, *An Ocean of Air: A Natural History of the Atmosphere* (London: Bloomsbury, 2010).

28. Vaclav Smil, "Harvesting the Biosphere: The Human Impact," *Population and Development Review* 37, no. 4 (2011): 613–636.

29. Paul Jepson and Richard J. Ladle, "Nature Apps: Waiting for the Revolution," *Ambio* 44, no. 8 (2015): 827–832.

30. Louv, *Last Child in the Woods.*

31. Jamil Zaki, "Moving beyond Stereotypes of Empathy," *Trends in Cognitive Sciences* 21, no. 2 (2017): 59–60.

32. Sarah H. Konrath, Edward H. O'Brien, and Courtney Hsing, "Changes in Dispositional Empathy in American College Students over Time: A Meta-Analysis," *Personality and Social Psychology Review* 15, no. 2 (2011): 180–198.

33. Adrian David Cheok and Emma Yann Zhang, "Kissenger: Transmitting Kiss through the Internet," in *Human–Robot Intimate Relationships* (Cham: Springer, 2019), 77–97.

34. Liora Gubkin, "From Empathetic Understanding to Engaged Witnessing: Encountering Trauma in the Holocaust Classroom," *Teaching Theology & Religion* 18, no. 2 (2015): 103–120.

35. Keziah Wallis and Miriam Ross, "Fourth VR: Indigenous Virtual Reality Practice," *Convergence* 27, no. 2 (2021): 313–329.

36. Loretta Todd, "Aboriginal Narratives in Cyberspace," in *Immersed in Technology: Art and Virtual Environments*, ed. Mary Anne Moser and Douglas Macleod (Cambridge, MA: MIT Press, 1996), 179–194.

37. Jackson 2Bears, "A Conversation with Spirits inside the Simulation of a Coast Salish Longhouse," in *Code Drift: Essays in Critical Digital Studies*, ed. Arthur Kroker and Marilouise Kroker (Victoria, BC: New World Perspectives/CTheory Books, 2010), 153–175.

CHAPTER 9

1. Alexander Verderber, Michael McKnight, and Alper Bozkurt, "Early Metamorphic Insertion Technology for Insect Flight Behavior Monitoring," *Journal of Visualized Experiments* 89 (2014): 1–8.

2. Alper Bozkurt, Amit Lal, and Robert Gilmour, "Radio Control of Insects for Biobotic Domestication," in *2009 4th International IEEE/EMBS Conference on Neural Engineering*, Antalya, Turkey, February 23–March 13 (New York: IEEE, 2009), 215–218.

3. Kevin Dai, Ravesh Sukhnandan, Michael Bennington, Karen Whirley, Ryan Bao, Lu Li, Jeffrey P. Gill, et al., "SLUGBOT, an Aplysia-Inspired Robotic Grasper for Studying Control," in *Biomimetic and Biohybrid Systems: 11th International Conference, Living Machines 2022, Virtual Event, July 19–22, 2022, Proceedings* (Cham: Springer International, 2022), 182–194; Victoria A. Webster et al., "Organismal Engineering: Toward a Robotic Taxonomic Key for Devices Using Organic Materials," *Science Robotics* 2, no. 12 (2017): 1–18.

4. Alper Bozkurt, Edgar Lobaton, and Mihail Sichitiu, "A Biobotic Distributed Sensor Network for Under-Rubble Search and Rescue," *Computer* 49, no. 5 (2016): 38–46.

5. Josh Bongard, "From Rigid to Soft to Biological Robots: How New Materials Are Driving Advances in the Study of the Embodied Cognition," *Artificial Life and Robotics* 28, no. 2 (2023): 282–286; Rafael Mestre, Tania Patiño, and Samuel Sánchez, "Biohybrid Robotics: From the Nanoscale to the Macroscale," *Wiley Interdisciplinary Reviews: Nanomedicine and Nanobiotechnology* 13, no. 5 (2021): e1703; Zening Lin et al., "The Emerging Technology of Biohybrid Micro-Robots: A Review," *Bio-Design and Manufacturing* vol. 5 (2022): 107–132; Leonardo Ricotti and Arianna Menciassi, "Nanotechnology in Biorobotics: Opportunities and Challenges," *Journal of Nanoparticle Research* 17 (2015): 1–10.

6. Guangye Li and Dingguo Zhang, "Brain-Computer Interface Controlled Cyborg: Establishing a Functional Information Transfer Pathway from Human Brain to Cockroach Brain," *PLOS One* 11, no. 3 (2016): 1–17.

7. Andrés Iglesias, Akemi Gálvez, and Patricia Suárez, "Swarm Robotics—A Case Study: Bat Robotics," in *Nature-Inspired Computation and Swarm Intelligence: Algorithms, Theory and Applications*, ed. Xin-She Yang (London: Academic Press, 2020), 273–302.

8. Yao Li and Hirotaka Sato, "Insect-Computer Hybrid Robot," *Molecular Frontiers Journal* 2, no. 1 (2018): 30–42.

9. Ki-Hun Jeong, Jaeyoun Kim, and Luke P. Lee, "Biologically Inspired Artificial Compound Eyes," *Science* 312, no. 5773 (2006): 557–561.

10. Donato Romano et al., "A Review of Animal-Robot Interaction: From Bio-Hybrid Organisms to Mixed Societies," *Biological Cybernetics* 113, no. 3 (2019): 1–28, 201–225; Elizabeth R. Johnson and Jesse Goldstein, "Biomimetic Futures: Life, Death, and the Enclosure of a More-Than-Human Intellect," *Annals of the Association of American Geographers* 105, no. 2 (2015): 387–396.

11. Feng Cao et al., "Insect-Computer Hybrid Legged Robot with User-Adjustable Speed, Step Length and Walking Gait," *Journal of the Royal Society Interface* 13, no. 116 (2016): 1–10; Guangye Li and Dingguo Zhang, "Brain-Computer Interface Controlling Cyborg: A Functional Brain-to-Brain Interface between Human and Cockroach," in *Brain-Computer Interface Research: A State-of-the Art Summary*, ed. Christoph Guger, Brendan Z. Allison, and Günter Edlinger (Cham: Springer, 2017), 71–79.

12. Tedrick Lew et al., "The Emergence of Plant Nanobionics and Living Plants as Technology," *Advanced Materials Technologies* 5, no. 3 (2020).

13. Tedrick Lew et al., "Real-Time Detection of Wound-Induced H_2O_2 Signalling Waves in Plants with Optical Nanosensors," *Nature Plants* 6, no. 4 (2020b): 404–415.

14. Anne Trafton, "Bionic Plants," *MIT News*, March 16, 2014, http://news.mit.edu/2014/bionic-plants.

15. Ardemis A. Boghossian et al., "Application of Nanoparticle Antioxidants to Enable Hyperstable Chloroplasts for Solar Energy Harvesting," *Advanced Energy Materials* 3, no. 7 (2013): 881–893.

16. Daniela Rus and Michael T. Tolley, "Design, Fabrication and Control of Soft Robots," *Nature* 521, no. 7553 (2015): 467–475.

17. Sang Jin Kim et al., "Materials for Flexible, Stretchable Electronics: Graphene and 2D Materials," *Annual Review of Materials Research* 45 (2015): 63–84; Christina Larson et al., "Highly Stretchable Electroluminescent Skin for Optical Signaling and Tactile Sensing," *Science* 351, no. 6277 (2016): 1071–1074.

18. Cecilia Laschi, Barbara Mazzolai, and Matteo Cianchetti, "Soft Robotics: Technologies and Systems Pushing the Boundaries of Robot Abilities," *Science Robotics* 1, no. 1 (2016): 1–11; Carmel Majidi, "Soft-Matter Engineering for Soft Robotics," *Advanced Materials Technologies* 4, no. 2 (2019): 1–13.

19. Rika Wright Carlsen and Metin Sitti, "Bio-Hybrid Cell-Based Actuators for Microsystems," *Small* 10, no. 19 (2014): 3831–3851.

20. Leonardo Ricotti et al., "Biohybrid Actuators for Robotics: A Review of Devices Actuated by Living Cells," *Science Robotics* 2, no. 12 (2017): 1–17.

21. Sau Yin Chin et al., "Additive Manufacturing of Hydrogel-Based Materials for Next-Generation Implantable Medical Devices," *Science Robotics* 2, no. 2 (2017): 1–10.

22. Veronika Magdanz, Samuel Sanchez, and Oliver G. Schmidt, "Development of a Sperm-Flagella Driven Micro-Bio-Robot," *Advanced Materials* 25, no. 45 (2013): 6581–6588.

23. Melanie J. Anderson et al., "A Bio-Hybrid Odor-Guided Autonomous Palm-Sized Air Vehicle," *Bioinspiration & Biomimetics* 16, no. 2 (2020): 026002.

24. Francisco Sánchez-Bayo and Kris A. G. Wyckhuys, "Worldwide Decline of the Entomofauna: A Review of Its Drivers," *Biological Conservation* 232 (2019): 8–27.

25. Sam Kriegman et al., "A Scalable Pipeline for Designing Reconfigurable Organisms," *Proceedings of the National Academy of Sciences* 117, no. 4 (2020): 1853–1859; Douglas Blackiston et al., "A Cellular Platform for the Development of Synthetic Living Machines," *Science Robotics* 6, no. 52 (2021): eabf1571.

26. "Xenobot," *GitHub*, accessed January 19, 2020, https://github.com/xenobot-dev/xenobot.

27. Jitka Čejková et al., "Droplets as Liquid Robots," *Artificial Life* 23, no. 4 (2017): 528–549; Janna C. Nawroth et al., "A Tissue-Engineered Jellyfish with Biomimetic Propulsion," *Nature Biotechnology* 30, no. 8 (2012): 792–797.

28. Masahiro Mori, "The Uncanny Valley: The Original Essay by Masahiro Mori," *IEEE Spectrum*, June 12, 2012.

29. Adrienne Mayor, *Gods and Robots: Myths, Machines, and Ancient Dreams of Technology* (Princeton, NJ: Princeton University Press, 2020).

30. Amy Hinterberger, "Regulating Estrangement: Human-Animal Chimeras in Postgenomic Biology," *Science, Technology, and Human Values* 45, no. 6 (2020): 1065–1086; Brian Salter and Alison Harvey, "Creating Problems in the Governance of Science: Bioethics and Human/Animal Chimeras," *Science and Public Policy* 41, no. 5 (2014): 685–696.

31. Alison Harvey and Brian Salter, "Anticipatory Governance: Bioethical Expertise for Human/Animal Chimeras," *Science as Culture* 21, no. 3 (2012): 291–313.

32. Douglas Blackiston et al., "Biological Robots: Perspectives on an Emerging Interdisciplinary Field," *Soft Robotics*, April 20, 2023.

33. David Sepkoski, "'Replaying Life's Tape': Simulations, Metaphors, and Historicity in Stephen Jay Gould's View of Life," *Studies in History and Philosophy of Science Part C: Studies in History and Philosophy of Biological and Biomedical Sciences* 58 (2016): 73–81.

34. Lynn Sagan, "On the Origin of Mitosing Cells," *Journal of Theoretical Biology* 14, no. 3 (1967).

35. Lynn Margulis, *Symbiosis in Cell Evolution: Life and Its Environment on the Early Earth* (San Francisco: W. H. Freeman, 1981).

36. Scott F. Gilbert, Jan Sapp, and Alfred I. Tauber, "A Symbiotic View of Life: We Have Never Been Individuals," *The Quarterly Review of Biology* 87, no. 4 (2012): 325–341; Margaret McFall-Ngai et al., "Animals in a Bacterial World, a New Imperative for the Life Sciences," *Proceedings of the National Academy of Sciences* 110, no. 9 (2013): 3229–3236.

37. J. E. Lovelock and L. Margulis, "Atmospheric Homeostasis by and for the Biosphere: The Gaia Hypothesis," *Tellus* 26, nos. 1–2 (1974): 2–10; L. Margulis et al., *Scientists Debate Gaia: The Next Century* (Cambridge, MA: MIT Press, 2004).

CHAPTER 10

1. Hyojin Kim, Daniel Bojar, and Martin Fussenegger, "A CRISPR/Cas9-Based Central Processing Unit to Program Complex Logic Computation in Human Cells," *Proceedings of the National Academy of Sciences* 116, no. 15 (2019): 7214–7219.

2. Julius Fredens et al., "Total Synthesis of *Escherichia coli* with a Recoded Genome," *Nature* 569, no. 7757 (2019): 514–518.

3. Van Chinh Tran et al., "Electrical Current Modulation in Wood Electrochemical Transistor," *Proceedings of the National Academy of Sciences* 120, no. 18 (2023): e2218380120.

4. Ethan Bier, "Gene Drives Gaining Speed," *Nature Reviews Genetics* 23, no. 1 (2022): 5–22; Kevin M. Esvelt and Neil J. Gemmell, "Conservation Demands Safe Gene Drive," *PLOS Biology* 15, no. 11 (2017): 1–8; Kenneth A. Oye et al., "Regulating Gene Drives," *Science* 345, no. 6197 (2014): 626–628.

5. Syed Shan-e-Ali Zaidi, Ahmed Mahas, Hervé Vanderschuren, and Magdy M. Mahfouz, "Engineering Crops of the Future: CRISPR Approaches to Develop Climate-Resilient and Disease-Resistant Plants," *Genome Biology* 21, no. 1 (2020): 1–19.

6. Aftab Ahmad et al., "CRISPR/Cas-Mediated Abiotic Stress Tolerance in Crops," in *CRISPR Crops: The Future of Food Security*, ed. Aftab Ahmad, Sultan Habibullah Khan, and Zulqurnain Khan (Singapore: Springer, 2021), 177–211; Akshaya K. Biswal et al., "CRISPR

Mediated Genome Engineering to Develop Climate Smart Rice: Challenges and Opportunities," *Seminars in Cell & Developmental Biology* 96 (2019): 100–106.

7. Ken Anthony et al., "New Interventions Are Needed to Save Coral Reefs," *Nature Ecology & Evolution* 1, no. 10 (2017): 1420–1422; Philip A. Cleves et al., "Reduced Thermal Tolerance in a Coral Carrying CRISPR-Induced Mutations in the Gene for a Heat-Shock Transcription Factor," *Proceedings of the National Academy of Sciences* 117, no. 46 (2020): 28899–28905; Jeroen A. J. M. van de Water et al., "Coral Holobionts and Biotechnology: From Blue Economy to Coral Reef Conservation," *Current Opinion in Biotechnology* 74 (2022): 110–121; Madeleine J. H. van Oppen and John G. Oakeshott, "A Breakthrough in Understanding the Molecular Basis of Coral Heat Tolerance," *Proceedings of the National Academy of Sciences* 117, no. 46 (2020): 28546–28548.

8. Robert Pollack, "Eugenics Lurk in the Shadow of CRISPR," *Science* 348, no. 6237 (2015): 871.

9. Gregory D. Scholes and Edward H. Sargent, "Boosting Plant Biology," *Nature Materials* 13, no. 4 (2014): 329–331; Juan Pablo Giraldo et al., "Plant Nanobionics Approach to Augment Photosynthesis and Biochemical Sensing," *Nature Materials* 13, no. 4 (2014): 400–408.

10. Seon-Yeong Kwak et al., "A Nanobionic Light-Emitting Plant," *Nano Letters* 17, no. 12 (2017): 7951–7961.

11. Martin Röck et al., "Embodied GHG Emissions of Buildings: The Hidden Challenge for Effective Climate Change Mitigation," *Applied Energy* 258 (2020): 114107.

12. Mary Katherine Heinrich et al., "Constructing Living Buildings: A Review of Relevant Technologies for a Novel Application of Biohybrid Robotics," *Journal of the Royal Society Interface* 16, no. 156 (2019): 20190238; Mette Ramsgaard Thomsen and Martin Tamke, "Towards a Transformational Eco-Metabolistic Bio-Based Design Framework in Architecture," *Bioinspiration & Biomimetics* 17, no. 4 (2022): 045005.

13. Phil Ayres et al., "Living Weaves: Steps towards the Persistent Modelling of Bio-Hybrid Architectural Systems," in *Impact: Design with All Senses: Proceedings of the Design Modelling Symposium 2019*, Berlin, ed. Christoph Gengnagel, Olivier Baverel, Jane Burry, Mette Ramsgaard Thomsen, and Stefan Weinzierl (Cham: Springer, 2019), 446–459.

14. Phil Ayres, "Flora Robotica: Investigating a Living Bio-Hybrid Architecture," in *Living Architecture Systems Group Symposium 2019*, ed. Philip Beesley and Sascha Hastings (Cambridge, ON: Riverside Architectural Press, 2019), 83–86.

15. Heiko Hamman et al., "*Flora Robotica*: An Architectural System Combining Living Natural Plants and Distributed Robots," *arXiv:1709.04291* (2017).

16. Jacqueline T. Chien et al., "Biogotchi! An Exploration of Plant-Based Information Displays," in *Proceedings of the 33rd Annual ACM Conference Extended Abstracts on Human Factors in Computing Systems*, Seoul, Republic of Korea (New York: Association for Computing Machinery, 2015), 1139–1144; Vito Gentile et al., "Plantxel: Towards a Plant-Based Controllable Display," in *Proceedings of the 7th ACM International Symposium on Pervasive Displays*,

Munich, Germany, June 6–8 (New York: Association for Computing Machinery, 2018), 1–8; David Holstius et al., "Infotropism: Living and Robotic Plants as Interactive Displays," in *Proceedings of the 5th Conference on Designing Interactive Systems: Processes, Practices, Methods, and Techniques*, Cambridge, MA, August 1–4 (New York: Association for Computing Machinery, 2004), 215–221.

17. Tedrick Lew et al., "The Emergence of Plant Nanobionics and Living Plants as Technology," *Advanced Materials Technologies* 5, no. 3 (2020): 1900657.
18. Tetsuo Tsutsui and Katsuhiko Fujita, "The Shift from 'Hard' to 'Soft' Electronics," *Advanced Materials* 14, no. 13–14 (2002): 949–952.
19. Itsuki Kunita et al., "Adaptive Path-Finding and Transport Network Formation by the Amoeba-Like Organism *Physarum*," in *Natural Computing and Beyond*, ed. Yasuhiro Suzuki and Toshiyuki Nakagaki (Tokyo: Springer, 2013), 14–29; Atsushi Tero et al., "A Method Inspired by *Physarum* for Solving the Steiner Problem," *International Journal of Unconventional Computing* 6, no. 2 (2010).
20. Merlin Sheldrake, *Entangled Life: How Fungi Make Our Worlds, Change Our Minds and Shape Our Futures* (New York: Random House, 2020).
21. Mihai Irimia-Vladu et al., "Biocompatible and Biodegradable Materials for Organic Field-Effect Transistors," *Advanced Functional Materials* 20, no. 23 (2010): 4069–4076; Hu Tao et al., "Silk-Based Conformal, Adhesive, Edible Food Sensors," *Advanced Materials* 24, no. 8 (2012): 1067–1072; Ziyi Guo et al., "Nanobiohybrids: Materials Approaches for Bioaugmentation," *Science Advances* 6, no. 12 (2020): 1–16; Evgeny Katz, "Biocomputing: Tools, Aims, Perspectives," *Current Opinion in Biotechnology* 34 (2015): 202–208.
22. Sang Yup Lee et al., "DNA Data Storage Is Closer Than You Think," *Scientific American*, July 1, 2019, https://www.scientificamerican.com/article/dna-data-storage-is-closer-than-you-think/.
23. Seth L. Shipman et al., "CRISPR–Cas Encoding of a Digital Movie into the Genomes of a Population of Living Bacteria," *Nature* 547, no. 7663 (2017): 345–349.
24. Christopher N. Takahashi et al., "Demonstration of End-to-End Automation of DNA Data Storage," *Nature Scientific Reports* 9, no. 1 (2019): 4998.
25. Victor Zhirnov et al., "Nucleic Acid Memory," *Nature Materials* 15, no. 4 (2016): 366–370.
26. Victor D. Chase, "Team Develops Biological Computer," *Nature Medicine* 5, no. 722 (1999).
27. Andrew Adamatzky et al. "Fungal Electronics," *Biosystems* 212 (2022): 104588; Andrew Adamatzky, "Towards Fungal Computer," *Interface Focus* 8, no. 6 (2018): 1–6.
28. Sheldrake, *Entangled Life.*
29. Ivan Poupyrev, Chris Harrison, and Munehiko Sato, "Touché: Touch and Gesture Sensing for the Real World," in *Proceedings of the 2012 ACM Conference on Ubiquitous Computing*,

Pittsburgh, PA, September 5–8 (New York: Association for Computing Machinery, 2012), 536.

30. Ivan Poupyrev, "Botanicus Interacticus," *Ivan Poupyrev*, 2012, http://www.ivanpoupyrev.com/project/botanicus-interacticus.

31. Mark Weiser, "The Computer for the 21st Century," *Scientific American* 265, no. 3 (1991): 94–105.

32. René Doursat, Hiroki Sayama, and Olivier Michel, "A Review of Morphogenetic Engineering," *Natural Computing* 12, no. 4 (2013): 517–535; Seth Copen Goldstein et al., "Programmable Matter," *Computer* 38, no. 6 (2005): 99–101.

Index